AS/A-LEVEL YEAR 2

STUDENT GUIDE

EDEXCEL

Physics

Topics 6–8

Further mechanics

Electric and magnetic fields

Nuclear and particle physics

Mike Benn

PHILIP ALLAN FOR
HODDER
EDUCATION
AN HACHETTE UK COMPANY

Special thanks to Graham George for his help and advice in the publication of this book.

Philip Allan, an imprint of Hodder Education, an Hachette UK company, Blenheim Court, George Street, Banbury, Oxfordshire OX16 5BH

Orders

Bookpoint Ltd, 130 Park Drive, Milton Park, Abingdon, Oxfordshire OX14 4SE

tel: 01235 827827

fax: 01235 400401

e-mail: education@bookpoint.co.uk

Lines are open 9.00 a.m.–5.00 p.m., Monday to Saturday, with a 24-hour message answering service. You can also order through the Hodder Education website: www.hoddereducation.co.uk

This guide has been written specifically to support students preparing for the Edexcel A-level physics examinations. The content has been neither approved nor endorsed by Edexcel and remains the sole responsibility of the author.

Cover photo: alexskopje/Fotolia

Typeset by Integra Software Services Pvt. Ltd, Pondicherry, India

Printed in Italy

Hachette UK's policy is to use papers that are natural, renewable and recyclable products and made from wood grown in sustainable forests. The logging and manufacturing processes are expected to conform to the environmental regulations of the country of origin.

Contents

Getting the most from this book 4

About this book 5

Content Guidance

Further mechanics 6

    Momentum and impulse 6

    Newton's second law of motion 6

    The conservation of linear momentum 8

    Collisions 10

    Circular motion 13

Electric and magnetic fields 16

    Electric fields 16

    Capacitance 22

    Magnetic fields 27

Nuclear and particle physics 35

    The nuclear atom 35

    Particle accelerators 37

    Particle detectors 41

    Particle interactions 42

    The quark–lepton model 45

Questions & Answers

Test paper 1 51

Test paper 2 66

Knowledge check answers 84

Index 86

■Getting the most from this book

Exam-style questions

Commentary on the questions

Tips on what you need to do to gain full marks, indicated by the icon **e**

Sample student answers

Practise the questions, then look at the student answers that follow.

Commentary on sample student answers

Find out how many marks each answer would be awarded in the exam and then read the comments (preceded by the icon **e**) following each student answer showing exactly how and where marks are gained or lost.

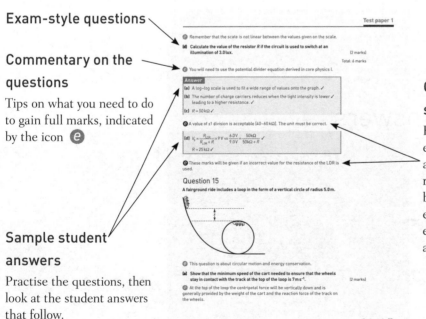

■About this book

This guide is one of a series covering the Edexcel specification for AS and A-level physics. It offers advice for the effective study of Topic 6 (Further mechanics), Topic 7 (Electric and magnetic fields) and Topic 8 (Nuclear and particle physics). Its aim is to help you *understand* the physics — it is not intended as a shopping list, enabling you to cram for the examination. The guide has two sections:

- The **Content Guidance** is not intended to be a detailed textbook. It offers guidance on the main areas of the content of the topics, with an emphasis on worked examples. These examples illustrate the types of question that you are likely to come across in the examination.

- The **Questions & Answers** section comprises two sample tests — one with the structure and style of the A-level paper 1 examination, the other with the structure of an A-level paper 3. Although most of the questions will be restricted to the topics in this guide, you are expected to have studied the topics of the core physics section and there will be some questions or part questions on topics 2 and 3 (Mechanics and Electric circuits) in test paper 1, and in test paper 2 there may be some synoptic questions requiring a knowledge of all the core physics topics. Answers are provided and, in some cases, distinction is made between responses that might have been written by an A-grade student and those typical of a C-grade student. Common errors made by students are also highlighted so that you, hopefully, do not make the same mistakes.

The purpose of this book is to help you with the A-level papers 1 and 3, but don't forget that what you are doing is learning physics. The development of an understanding of physics can only evolve with experience, which means time spent thinking about physics, working with it and solving problems. This book provides you with a platform for doing this.

If you try all the worked examples and the tests before looking at the answers, you will begin to think for yourself and develop the necessary techniques for answering examination questions. In addition, you will need to learn the basic formulae, definitions and experiments. Thus prepared, you will be able to approach the examination with confidence.

The specification states the physics that will be examined in the A-level examinations and describes the format of those tests. This is not necessarily the same as what teachers might choose to teach (or what you might choose to learn).

The specification can be obtained from Edexcel, either as a printed document or from the web at www.edexcel.com.

Content Guidance

■ Further mechanics

This section follows on from the Mechanics section of the core physics covered in the first of this series of guides. You should be familiar with the concepts of momentum and Newton's laws of motion, but some of the basic material will be repeated here.

Momentum and impulse

Momentum, p, is the product of the mass and velocity of an object. It is a vector quantity and has the unit $kg\,m\,s^{-1}$.

An **impulse** occurs when a body experiences a sudden, short force — such as that applied to a tennis ball when it is struck by a racquet.

In most cases, like that of the tennis ball, the force applied is not constant but varies as shown in Figure 1.

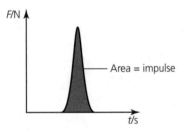

Figure 1

The impulse is represented by the sum of the instantaneous impulses ($\int F\,dt$), which is equal to the area under the graph.

Newton's second law of motion

In the first guide in this series, which covers topics 2 and 3, you used a version of **Newton's second law of motion** that applies only to the special case where a resultant force acts on a body of fixed mass and causes the body to accelerate. The law can be stated more generally in terms of the change in momentum that occurs when a resultant force is applied to a body.

In symbols:

$$\Sigma F = \frac{\Delta p}{\Delta t}$$

(if consistent SI units are used)

This statement of Newton's second law covers situations in which the mass is indeterminate (e.g. interactions involving neutrinos) or continually changing

momentum =
                    mass × velocity

$p = mv$

impulse $= F\Delta t = \Delta p$

Knowledge check 1

Calculate the momentum of a tennis ball of mass 50 g moving with a velocity of $40\,m\,s^{-1}$.

Knowledge check 2

Calculate the impulse applied by the floor when a falling ball of mass 200 g strikes the floor when moving with a velocity of $12\,m\,s^{-1}$ and rebounds with a velocity of $8.0\,m\,s^{-1}$.

Newton's second law of motion states that the rate of change in momentum is directly proportional to the resultant applied force.

(e.g. a jet of water striking a surface). However, in paper 1 Newton's second law will be applied only in situations where the mass is constant.

For a fixed mass:

$$\sum F = \frac{\Delta(mv)}{\Delta t} = m\frac{\Delta v}{\Delta t} = ma$$

The above calculation confirms the validity of the equation used to represent Newton's second law when a force is applied to a fixed mass.

Core practical 9

Core practical 9 requires you to investigate the relationship between the force exerted on an object and its change in momentum. Full details of the measurements made, the precautions taken and how the results were analysed are needed.

Worked example

In an experiment to investigate Newton's second law, a glider is pulled along an air track by a falling mass, as shown in Figure 2.

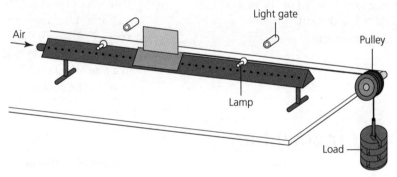

Figure 2

The gates are interfaced with a computer that displays the intervals when the interrupter card crosses the light gates as pulses. The results for one such experiment are shown as a computer display below.

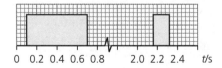

The mass of the glider is 400 g and the length of the interrupter card is 20.0 cm.

Use the results to calculate:

a The initial and final momentum of the glider.

b The resultant force acting on the glider.

→

The mass of the load was 22 g.

c i Calculate the total force acting on the system.

ii Explain why this is larger than the force acting on the glider.

d Calculate the acceleration of the glider when a load of 44 g is applied.

Answer

a initial velocity = $\dfrac{0.200\,\text{m}}{0.60\,\text{s}} = 0.333\,\text{m s}^{-1}$

initial momentum = $0.400\,\text{kg} \times 0.333\,\text{m s}^{-1} = 0.133\,\text{kg m s}^{-1}$

final velocity = $\dfrac{0.200\,\text{m}}{0.16\,\text{s}} = 1.25\,\text{m s}^{-1}$

final momentum = $0.400\,\text{kg} \times 1.25\,\text{m s}^{-1} = 0.500\,\text{kg m s}^{-1}$

b time between the light gates = $2.24\,\text{s} - 0.40\,\text{s} = 1.84\,\text{s}$

resultant force = $\dfrac{\text{change in momentum}}{\text{time}} = \dfrac{(0.500 - 0.133)\text{kg m s}^{-1}}{1.84\,\text{s}} = 0.20\,\text{N}$

c i The total force pulling the glider is the weight of the load, so:

$0.0220\,\text{kg} \times 9.81\,\text{m s}^{-2} = 0.22\,\text{N}$

ii The weight of the load (0.22 N) acts on the whole system, i.e. the mass of the glider plus that of the load. The glider is pulled by the tension in the string (0.22 N), which, in this case, is 0.02 N less than the weight.

d weight of load = (mass of glider + load) × acceleration

$a = \dfrac{0.044\,\text{kg} \times 9.81\,\text{m s}^{-2}}{(0.400 + 0.044)\text{kg}} = 0.97\,\text{m s}^{-2}$

The conservation of linear momentum

The **conservation of linear momentum** is a fundamental law of physics. The application of the principle was covered in the first student guide in this series covering Topics 2 and 3, but was restricted to interactions that occurred along a straight line. Later in this guide the law will be applied to interactions of subatomic particles, where non-linear events occur.

It is important that you fully understand the meaning of the expression 'system of interacting bodies'. It is evident that if two snooker balls collide and one of the balls is slowed down while the other speeds up, the momentum of both balls will change. However, the *vector sum* of the momentums of the balls after the collision will be the same as that immediately before the interaction.

The 'system' for a bouncing ball is more difficult to imagine. The Earth is the second body in the system, but it is so massive compared to the ball that the change in its momentum due to the interaction will be imperceptible — nonetheless, the principle of conservation of momentum still applies.

> **Exam tip**
>
> It is important to realise that this law does not contradict the law of conservation of momentum. Although the application of a resultant force may change the momentum of individual masses, the total momentum change for a system of two or more masses will always be zero.

> **Knowledge check 3**
>
> Calculate the average force exerted by a drop of water of mass 0.20 g falling from rest at a height of 1.0 m onto a surface. Assume that the water comes to rest in a time of 80 ms after striking the surface.

> The **conservation of linear momentum** states that in any system of interacting bodies, the total momentum is conserved, provided that no resultant external force acts on the system.

Worked example

Two gliders moving towards each other on an air track collide and then bounce apart, as shown in Figure 3. Initially the 400 g glider travels from left to right with a velocity of $0.80\,\mathrm{m\,s^{-1}}$ and the 200 g glider moves in the opposite direction at $0.40\,\mathrm{m\,s^{-1}}$. After the collision the smaller glider changes direction and moves back with a velocity of $0.80\,\mathrm{m\,s^{-1}}$.

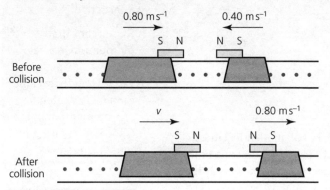

Figure 3

a Calculate the velocity of the larger glider after the collision.

b If the gliders stuck together on impact, calculate the velocity of the combination.

Answer

a Let the rightward direction correspond to positive values, and let v denote the velocity of the larger glider after the collision. Then, for this system:

$$\text{initial momentum} = 0.400\,\mathrm{kg} \times 0.80\,\mathrm{m\,s^{-1}} + 0.200\,\mathrm{kg} \times (-0.40\,\mathrm{m\,s^{-1}})$$
$$= 0.240\,\mathrm{kg\,m\,s^{-1}}$$
$$\text{final momentum} = 0.400\,\mathrm{kg} \times v + 0.200\,\mathrm{kg} \times 0.80\,\mathrm{m\,s^{-1}}$$

Conservation of momentum gives:

$$0.240\,\mathrm{kg\,m\,s^{-1}} = 0.400\,\mathrm{kg} \times v + 0.200\,\mathrm{kg} \times 0.80\,\mathrm{m\,s^{-1}}$$
so $v = +0.20\,\mathrm{m\,s^{-1}}$

Thus the larger glider moves with a velocity of $0.20\,\mathrm{m\,s^{-1}}$ to the right.

b $0.240\,\mathrm{kg\,m\,s^{-1}} = (0.400\,\mathrm{kg} + 0.200\,\mathrm{kg}) \times v_{\mathrm{comb}}$

$$v_{\mathrm{comb}} = +0.40\,\mathrm{m\,s^{-1}}$$

The combination of two gliders moves with a velocity of $0.40\,\mathrm{m\,s^{-1}}$ to the right.

For oblique collisions, in which the masses approach or separate along different lines, the principle of conservation of momentum can be applied to the *components* of the momentum in any direction.

Worked example

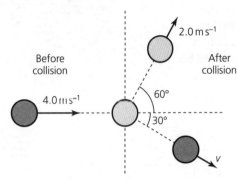

Figure 4

A snooker ball travelling at $4.0\,\text{m s}^{-1}$ strikes a stationary ball of identical mass, causing it to move with speed $2.0\,\text{m s}^{-1}$ at an angle of $60°$ to the direction of the incident ball, as shown in Figure 4. Calculate the velocity of the incident ball after the collision, given that it is deviated by $30°$ from its original path.

Answer

Let m kg be the mass of each ball. Taking components along the direction in which the snooker ball travels before the collision:

$$\text{initial momentum} = (m\,\text{kg} \times 4.0\,\text{m s}^{-1}) + (m\,\text{kg} \times 0\,\text{m s}^{-1}) = 4.0m\,\text{kg m s}^{-1}$$

$$\text{final momentum} = (m\,\text{kg} \times v\cos 30°) + (m\,\text{kg} \times 2.0\,\text{m s}^{-1}\cos 60°)$$

Because momentum is conserved:

$$4.0m\,\text{kg m s}^{-1} = (mv\cos 30°\,\text{kg}) + (2.0m\cos 60°\,\text{kg m s}^{-1})$$

The masses m cancel, giving $v = 3.5\ \text{m s}^{-1}$

Core practical 10

Core practical 10 requires you to use ICT to analyse collisions between small spheres, e.g. ball bearings on a table top. You will need to describe the details of how ICT is used to analyse images from video frames or stroboscopic photographs to explain how the principle of conservation of linear momentum can be applied to the system.

Collisions

The principles of **conservation of energy** and the conservation of momentum together form a cornerstone of physics at all levels.

The **conservation of energy** states that energy cannot be created or destroyed, although it can be transferred from one form to another.

Exam tip

It is useful to sketch a simple diagram showing the masses and velocities of bodies before and after an interaction.

Remember that momentum is a vector. If you assign positive values to left-to-right motion, the velocities and momentum in the right-to-left direction must be negative.

Knowledge check 4

A trolley of mass $2.50\,\text{kg}$ moving at $2.2\,\text{m s}^{-1}$ collides with a second trolley of mass $1.5\,\text{kg}$ that is travelling at $3.6\,\text{m s}^{-1}$ in the opposite direction. In the collision the trolleys stick together. Calculate the velocity of the trolleys after the collision (remember that velocity is a vector and both magnitude and direction are needed).

Exam tip

The same result can be achieved by taking components of the momentum at right angles to the direction of the incident ball, so that the total momentum is zero. Trying this would be a worthwhile exercise.

Both conservation principles must apply in collisions, but while the momentum of the system after the interaction is always exactly the same as it was before the interaction, energy can be transformed into other types. For example, if two identical cars moving in opposite directions at the same speed collide, they will both stop on impact and they appear to have lost both momentum and kinetic energy. In fact the initial momentum of the system was zero because the cars were travelling in opposite directions, and on impact the kinetic energy of the cars is transformed to other forms of energy, predominantly thermal energy.

Collisions are usually classified in terms of the *kinetic* energy in the system before and after the interaction.

- In **elastic collisions**, all of the kinetic energy in the system is conserved.
- In **inelastic collisions**, some or all of the kinetic energy is transferred to other forms of energy.

In everyday life most collisions are inelastic. However, for interactions involving subatomic particles, elastic collisions are not uncommon. It is also possible for there to be an increase in kinetic energy in a system — for example in explosions where some chemical energy is transformed to kinetic energy of the exploded fragments, and in nuclear radiation and nuclear fission where a mass defect is transferred as extra kinetic energy.

Worked examples

a Revisit the Worked example on p. 9 involving the collision of a 400 g glider and a 200 g glider. Use the same data to determine the kinetic energy of the gliders before and after the two given types of collision. Hence state whether each collision is elastic or inelastic.

b A proton of mass 1.7×10^{-27} kg moving at 2.0×10^6 m s^{-1} collides with a stationary alpha particle of mass 6.8×10^{-27} kg and rebounds directly backward. The alpha particle moves along the same line with a velocity of 8.0×10^5 m s^{-1}. Show that the collision is elastic.

Answers

a In the scenario where the gliders rebound:

$$\text{initial kinetic energy} = \tfrac{1}{2} \times 0.400\,\text{kg} \times (0.80\,\text{m s}^{-1})^2 + \tfrac{1}{2} \times 0.200\,\text{kg} \times (0.40\,\text{m s}^{-1})^2$$

$$= 0.144\,\text{J}$$

$$\text{final kinetic energy} = \tfrac{1}{2} \times 0.400\,\text{kg} \times (0.20\,\text{m s}^{-1})^2 + \tfrac{1}{2} \times 0.200\,\text{kg} \times (0.80\,\text{m s}^{-1})^2$$

$$= 0.072\,\text{J}$$

The collision is inelastic because 0.072 J of kinetic energy has been transferred to other forms.

In the scenario where the gliders stick together:

$$\text{initial kinetic energy} = 0.144\,\text{J (as in the rebound case)}$$

$$\text{final kinetic energy} = \tfrac{1}{2} \times (0.400 + 0.200)\,\text{kg} \times (0.40\,\text{m s}^{-1})^2 = 0.048\,\text{J}$$

➔

Knowledge check 5

A ball of mass 0.40 kg moving at 2.0 m s^{-1} collides with a stationary ball of mass 0.20 kg. After the collision both balls move in the same direction — the 0.40 kg ball travelling at 1.0 m s^{-1} and the smaller ball at 2.0 m s^{-1}. Is the collision elastic or inelastic?

The collision is inelastic because 0.096 J of kinetic energy has been transferred to other forms.

b First, we use the law of conservation of momentum to determine the velocity of the rebounding proton (Figure 5).

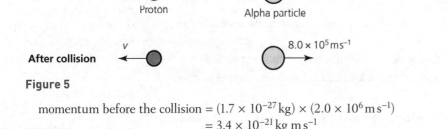

Figure 5

momentum before the collision $= (1.7 \times 10^{-27}\,\text{kg}) \times (2.0 \times 10^6\,\text{m s}^{-1})$
$$= 3.4 \times 10^{-21}\,\text{kg m s}^{-1}$$

momentum after the collision $= (1.7 \times 10^{-27}\,\text{kg} \times v) + (6.8 \times 10^{-27}\,\text{kg} \times 8.0 \times 10^5\,\text{m s}^{-1})$

Conservation of momentum gives:

$(1.7 \times 10^{-27}\,\text{kg} \times v) + (6.8 \times 8.0 \times 10^{-22}\,\text{kg m s}^{-1}) = 3.4 \times 10^{-21}\,\text{kg m s}^{-1}$

so $v = -1.2 \times 10^6\,\text{m s}^{-1}$

initial kinetic energy $= \frac{1}{2} \times (1.7 \times 10^{-27}\,\text{kg}) \times (2.0 \times 10^6\,\text{m s}^{-1})^2 = 3.4 \times 10^{-15}\,\text{J}$

final kinetic energy $= [\frac{1}{2} \times (1.7 \times 10^{-27}\,\text{kg}) \times (-1.2 \times 10^6\,\text{m s}^{-1})^2] +$
$$[\tfrac{1}{2} \times (6.8 \times 10^{-27}\,\text{kg}) \times (8.0 \times 10^5\,\text{m s}^{-1})^2]$$
$$= 3.4 \times 10^{-15}\,\text{J}$$

The kinetic energy is the same before and after the collision. So the collision is elastic.

Kinetic energy and momentum

It is often useful, particularly when studying collisions between subatomic particles, to convert momentum to kinetic energy and vice versa. We can derive an expression linking the two quantities $p = mv$ and $E_k = \frac{1}{2}mv^2$:

$p = mv \Rightarrow p^2 = m^2v^2$

so $\dfrac{p^2}{2m} = \dfrac{m^2v^2}{2m} = \dfrac{1}{2}mv^2$

and hence: $E_k = \dfrac{p^2}{2m}$

> **Exam tip**
>
> Momentum is a vector quantity, so the negative sign of v in Worked example (b) indicates that the proton travels in the opposite direction after the collision. Kinetic energy is a scalar quantity so the direction is irrelevant.

Worked examples

a Calculate the kinetic energy of an electron moving with a momentum of $2.0 \times 10^{-23}\,\text{kg m s}^{-1}$.

b Calculate the momentum of a rifle bullet of mass $5.0\,\text{g}$ fired with kinetic energy $400\,\text{J}$.

Answers

a $E_k = \dfrac{p^2}{2m} = \dfrac{(2.0 \times 10^{-23}\,\text{kg m s}^{-1})^2}{2 \times 9.1 \times 10^{-31}\,\text{kg}} = 2.2 \times 10^{-16}\,\text{J}$

b $p = \sqrt{2mE_k} = \sqrt{2 \times (5.0 \times 10^{-3}\,\text{kg}) \times 400\,\text{J}} = 2.0\,\text{kg m s}^{-1}$

Circular motion

Consider an object moving in a circle of radius r. You will need to know the definitions of **angular displacement**, **angular speed**, **period** and **frequency** and the equations linking these quantities.

Angular displacement is measured in radians (rad):

2π radians $= 360°$

1 radian $\approx 57°$

angular displacement $\theta = \dfrac{\text{arc length}}{\text{radius}} = \dfrac{\Delta s}{r}$

angular speed $\omega = \dfrac{\Delta \theta}{\Delta t} = \dfrac{\Delta s / \Delta t}{r} = \dfrac{v}{r}$, where v is the linear speed of the object (and so $v = \omega r$)

period $T = \dfrac{2\pi}{\omega}$

frequency $f = \dfrac{1}{T} = \dfrac{\omega}{2\pi}$

Worked example

a Show that the angular speed at which the Moon orbits the Earth is approximately $3 \times 10^{-6}\,\text{rad s}^{-1}$.

b Assuming that the Moon has a circular orbit of radius $3.9 \times 10^8\,\text{m}$ around the Earth, calculate its linear speed.

Answer

a $\omega = \dfrac{2\pi}{T}$

where T, the time taken for the Moon to orbit the Earth once, is 28 days.

So $\omega = \dfrac{2\pi\,\text{rad}}{28 \times 24 \times 60 \times 60\,\text{s}} = 2.6 \times 10^{-6}\,\text{rad s}^{-1} \approx 3 \times 10^{-6}\,\text{rad s}^{-1}$

b $v = \omega r = (2.6 \times 10^{-6}\,\text{rad s}^{-1}) \times (3.9 \times 10^8\,\text{m}) = 1.0 \times 10^3\,\text{m s}^{-1}$

Exam tip

In particle physics, speeds often approach the speed of light ($c = 3.0 \times 10^8\,\text{m s}^{-1}$). However, be aware that the equation $E_k = p^2/2m$ can be applied only in *non-relativistic* situations, where the speed of the particles is less than about 10% of c. The mass of an electron used here is provided on the data sheet.

Knowledge check 6

A small object attached to a piece of string of length $0.60\,\text{m}$, is made to revolve along a circular path with a constant angular velocity. It sweeps out an angle of $114°$ in a time of $0.25\,\text{s}$. Calculate **a** the angular velocity of the string, **b** the period of the revolution, **c** the frequency of the revolutions.

Centripetal acceleration

Consider a body of mass m moving in a circle of radius r at constant speed v.

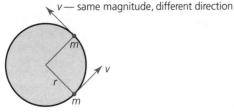

v — same magnitude, different direction

Figure 6

From Figure 6 you can see that while the magnitude of v is constant, its direction is continuously changing — in other words, although the linear speed is constant, the *velocity*, which is a vector quantity, is always changing.

- The change in velocity means that the mass is accelerating.
- Because the magnitude of v remains unchanged, the acceleration must always be directed at right angles to v, towards the centre of the circle.
- The magnitude of the acceleration is given by $a = v^2/r$.

This acceleration is known as **centripetal acceleration**.

Using the relationship $v = \omega r$, centripetal acceleration can also be expressed in terms of the angular speed:

$a = \omega^2 r$

By Newton's second law, a *resultant force* is needed to produce centripetal acceleration. This resultant force, called the **centripetal force**, is directed towards the centre of the circle and has a magnitude given by:

$$F = ma = \frac{mv^2}{r} = m\omega^2 r$$

Figure 7 shows the forces acting on a water skier.

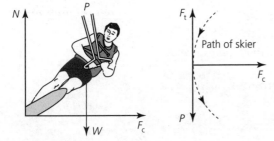

Figure 7

The tangential speed is constant so, by Newton's first law, the resultant tangential force must be zero. So the forward pull on the rope equals the backward drag of the water on the skis, i.e. $P = F_t$.

> **Exam tip**
>
> An object moving in a circle, or part of a circle, must be accelerating towards the centre. The velocity is a vector and because its direction is continuously changing towards the centre of the circle, it must accelerate in that direction.

> **Exam tip**
>
> In many calculations (such as satellite motion) the period is given and so $\omega = 2\pi/T$ can be used.

> **Knowledge check 7**
>
> Calculate the centripetal acceleration of the Moon assuming that it is in circular motion around the Earth with a period of 28 days at a radius of 4.0×10^8 m.

If the mass of the skier (plus board) is m, then we have:

$$F_c = \frac{mv^2}{r}$$

The centripetal force F_c comes from the sideways push of the water on the skis. So, according to Newton's third law, the sideways push of the skis on the water is $-F_c$.

Worked example

The satellite in Figure 8 orbits the Earth once every 87 minutes.

Figure 8

a Show that the angular speed of the satellite is approximately $1 \times 10^{-3}\,\mathrm{rad\,s^{-1}}$.

b Draw a free-body force diagram for the satellite when it is in the position shown.

c With reference to your free-body force diagram, explain why the satellite is accelerating.

d The radius of the satellite's orbit is 6500 km. Calculate the magnitude of its acceleration.

Answer

a $T = 87 \times 60$ seconds, so:

$$\omega = \frac{2\pi\,\mathrm{rad}}{87 \times 60\,\mathrm{s}} = 1.2 \times 10^{-3}\,\mathrm{rad\,s^{-1}} \approx 1 \times 10^{-3}\,\mathrm{rad\,s^{-1}}$$

b
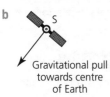
Gravitational pull
towards centre
of Earth

Figure 9

c The satellite is acted on by a single resultant force (Figure 9) that is directed towards the centre of the Earth; so it will accelerate in this direction.

d $a = \omega^2 r = (1.2 \times 10^{-3}\,\mathrm{rad\,s^{-1}})^2 \times (6500 \times 10^3\,\mathrm{m}) = 9.4\,\mathrm{m\,s^{-2}}$

In circular motion problems you should be aware of the cause of the centripetal force. In the example of the water skier, it is the push of the water against the skis that provides the resultant force; for the satellite, the gravitational pull of the Earth is what keeps it in orbit. Other sources of centripetal force include the tension in a string when a mass attached to the end of the string is whirled in a circle, and the frictional force of the road pushing on the tyres as a car rounds a bend.

Exam tip

Don't forget that, in a 'show that' answer, you should work out the answer to at least one more significant figure than in the given value, as evidence that you have actually worked through the calculation.

Summary

After studying this section, you should be able to:
- understand how to use the equation impulse $= F\Delta t = \Delta p$
- define Newton's second law of motion in terms of a change in momentum, and use the law in situations where the mass may not be constant
- understand how to apply the conservation of linear momentum to problems in two dimensions
- use the law of conservation of energy to describe a range of energy conversions, and to determine if a collision is elastic or inelastic
- derive and use the equation $E_k = p^2/2m$ for the kinetic energy of a non-relativistic particle
- understand that a body moving with constant speed in a circle is accelerating towards the centre of the circle
- use the equations $F = m\omega^2 r$ and $F = mv^2/r$ to work out the centripetal force in a variety of situations.

Electric and magnetic fields

In physics, a field represents a region in which certain objects will experience a force. In the first student guide in this series covering Topics 2 and 3 you studied some of the behaviour and properties of electromagnetic radiation, including the associated variation of electric fields and magnetic fields in waves. In this section, you look into the separate properties of these fields, and examine how the two are linked in the process of electromagnetic induction.

Electric fields

An **electric field** is generally represented by drawing **field lines,** or **lines of force**, which show the direction of the force that a very small positive charge would be subject to if it were placed in the field (see Figure 10).

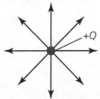

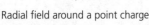

Radial field around a point charge   Uniform field between charged plates

Figure 10

The field around a point charge is 'radial', with field lines emanating like the spokes of a wheel. Between a pair of parallel conducting plates that have a potential difference applied across them the field is 'uniform', with all field lines perpendicular to the plates. We can use these diagrams to assess the relative field strengths at different places in the field. For example, near the point charge the field lines are more concentrated and so the field is strong; further from the point charge the lines spread out, indicating a weakening of the field strength; between the plates the lines of force are parallel and equally spaced, and the field strength is the same at all positions.

An **electric field** is a region in which a charged object will experience a force.

Exam tip

Many students lose marks in examinations by drawing sloppy, freehand diagrams. Always use a ruler to draw the field lines and ensure that radial lines are at the same angular separation, that uniform field lines are equally spaced and that arrows are included to show the direction of the field.

Electric field strength

An electric field has units $N\,C^{-1}$ and a strength given by:

$$E = \frac{F}{Q}$$

where F is the force that would act on a positive charge Q placed in the electric field.

E is a *vector* quantity whose direction is that of the force F. A useful expression for electric field strength is:

$$E = -\frac{dV}{dx} \qquad \text{unit: } V\,m^{-1}$$

Worked example

An electron gun in a cathode-ray tube has an electric field of strength $2.5 \times 10^5\,N\,C^{-1}$ between the cathode and the anode. Electrons from the heated cathode are accelerated across the 2.0 cm gap between the electrodes. Some pass through the hole in the anode and travel at constant speed until they strike a fluorescent screen (Figure 11).

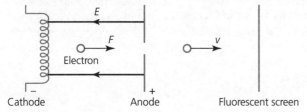

Figure 11

a Calculate the force on an electron in the field.
b Calculate the acceleration of the electron when it is between the electrodes.
c Assuming that the initial speed of the electron when it leaves the cathode is zero, calculate its speed when it leaves the gun.

Answer

a $F = EQ = (2.5 \times 10^5\,N\,C^{-1}) \times (1.6 \times 10^{-19}\,C) = 4.0 \times 10^{-14}\,N$

b $a = \dfrac{F}{m} = \dfrac{4.0 \times 10^{-14}\,N}{9.1 \times 10^{-31}\,kg} = 4.4 \times 10^{16}\,m\,s^{-2}$

c $v^2 = u^2 + 2as$

$= (0^2) + (2 \times 4.4 \times 10^{16}\,m\,s^{-2} \times 2.0 \times 10^{-2}\,m)$

So $v = 4.2 \times 10^7\,m\,s^{-1}$

Coulomb's law

In the section on DC electricity in core physics 1, you will have performed simple experiments with statically charged rods to demonstrate the nature of forces between charges:

■ like charges repel
■ unlike charges attract

Content Guidance

Coulomb's law shows how the magnitude of the electric force between two interacting charges depends on the magnitude of the charges and their distance apart:

$$F = k\frac{Q_1 Q_2}{r^2} = \frac{1}{4\pi\varepsilon_0}\frac{Q_1 Q_2}{r^2} = \frac{Q_1 Q_2}{4\pi\varepsilon_0 r^2}$$

where Q_1 and Q_2 are the magnitudes of the two point charges and r is the distance between them. The proportionality constant k is given by:

$$k = \frac{1}{4\pi\varepsilon_0}$$

where ε_0 is the **permittivity of free space** and has the value $8.85 \times 10^{-12}\,\text{F m}^{-1}$ (this is included in the data sheet at the end of each examination paper). In calculations the value $9.0 \times 10^9\,\text{N m}^2\,\text{C}^{-2}$ is often used for k (the more precise value $8.99 \times 10^9\,\text{N m}^2\,\text{C}^{-2}$ is given in the data sheet).

> **Coulomb's law** states that the force between two point charges is directly proportional to the product of the charges and inversely proportional to the square of their separation.

> **Knowledge check 9**
>
> Two small charged spheres exert a force of $1.8 \times 10^{-4}\,\text{N}$ on each other when they are 4 cm apart. What force will they exert when they are 12 cm apart?

Worked example

Two identical charged spheres are suspended from a point, as shown in Figure 12. Each sphere carries a charge of +50 nC, and the spheres are 10 cm apart. The threads are at an angle of 30°.

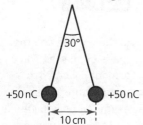

Figure 12

a Calculate the force between the two spheres.

b Draw a free-body force diagram for one of the spheres.

c Calculate the tension in each thread, and hence determine the mass of each sphere.

Answer

a $F = \dfrac{kQ_1 Q_2}{r^2} = \dfrac{(9.0 \times 10^9\,\text{N m}^2\,\text{C}^{-2}) \times (50 \times 10^{-9}\,\text{C}) \times (50 \times 10^{-9}\,\text{C})}{(10 \times 10^{-2}\,\text{m})^2}$

$= 2.25 \times 10^{-3}\,\text{N} \approx 2.3 \times 10^{-3}\,\text{N}$

b

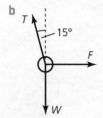

Figure 13

c Taking horizontal components:
$$T\sin 15° = 2.25 \times 10^{-3}\,\text{N so } T = \frac{2.25 \times 10^{-3}\,\text{N}}{\sin 15} = 8.7 \times 10^{-3}\,\text{N}$$

Taking vertical components:
$$mg = T\cos 15° \text{ so } m = \frac{8.7 \times 10^{-3}\,\text{N} \times \cos 15}{9.8\,\text{m s}^{-2}} = 8.6 \times 10^{-4}\,\text{kg} = 0.86\,\text{g}$$

The standard model of a hydrogen atom has an electron in a circular orbit around a proton. The centripetal force required to keep the electron in orbit is provided by the electrostatic force between the proton and the electron:

$$\frac{kQ_1Q_2}{r^2} = F = \frac{mv^2}{r}$$

The proton and the electron both have charge of magnitude $1.6 \times 10^{-19}\,\text{C}$; the mass of the electron is $9.1 \times 10^{-31}\,\text{kg}$ and the radius of a hydrogen atom is $0.11\,\text{nm}$.

Recall the diagram (Figure 10) of field lines for an electric field around a point charge. It was stated that the spreading out of the lines indicates that the field strength decreases with distance from the charge. An expression for the field strength around a point charge can be derived using Coulomb's law.

Consider the force F acting on a small charge δq at a distance r away from a point charge Q. We have:

$$F = \frac{Q\,\delta q}{4\pi\varepsilon_0 r^2} \text{ and } E = \frac{F}{\delta q}$$

It follows that:

$$E = \frac{Q}{4\pi\varepsilon_0 r^2}$$

So the field strength decreases with distance according to an inverse square law.

Electric potential

While electric field strength relates to forces on charged particles placed in the field, **electric potential** concerns the work done when a charge is moved in the field.

Within an electric field, lines or surfaces that are at the same potential are called **equipotentials**.

Because $E = dV/dx = 0$ along an equipotential, it follows that the equipotentials are always perpendicular to the electric field (where the component of E is zero) as shown in Figure 14.

Knowledge check 10

Using the data given in the text for the hydrogen atom, show that the speed of the orbiting electron is about $1.5 \times 10^6\,\text{m s}^{-1}$.

Electric potential, V is the work done per unit charge when a charged particle is moved in the field.

Equipotentials are lines or surfaces joining points of equal potential in an electric field.

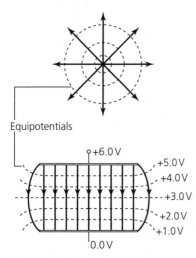

Equipotentials

+6.0V
+5.0V
+4.0V
+3.0V
+2.0V
+1.0V
0.0V

Figure 14

To determine the electric potential at a point, the work done per unit charge to move a small positive charge from that point to a position of zero potential must be evaluated.

To determine the value of the electric potential at a distance r from a point charge Q, you must calculate the work done per unit charge in moving a small positive charge from infinity (where the field due to Q is negligible and so the potential is zero) to that point. The work done can be found by integrating the force times the distance between infinity and the point.

$$W = \int_{\infty}^{r} F\,dx = \int_{\infty}^{r} E\,\delta q\,dx = \int_{\infty}^{r} \frac{Q\,\delta q}{4\pi\varepsilon_0 r^2}\,dx = \frac{Q\,\delta q}{4\pi\varepsilon_0 r^2}$$

$$V = \frac{W}{\delta q} = \frac{Q}{4\pi\varepsilon_0 r}$$

> **Exam tip**
>
> If the charge is negative, work would be done moving a small positive charge from the point to infinity (zero potential). So the potential at the point will be less than zero — it will have a negative value.

Worked example

a i What is the electric field strength at a distance of 50 cm from a point charge of +40 nC?

ii What would be the size of the force on a charge of +5.0 nC placed 50 cm from the +40 nC charge?

b i What is the potential in the field of a +4.0 μC point charge at a distance of 2.2 m from the charge?

ii A second charge of +2.0 μC is placed in the field 50 cm from the first charge. What is the potential energy in the system?

Answer

a i $E = \dfrac{Q}{4\pi\varepsilon_0 r^2} = \dfrac{40 \times 10^{-9}\,\text{C}}{4\pi(8.85 \times 10^{-12}\,\text{F m}^{-1}) \times (0.50\,\text{m})^2} = 1.4 \times 10^3\,\text{N C}^{-1}$

ii $F = EQ = 1.44 \times 10^3\,\text{N C}^{-1} \times 5.0 \times 10^{-9}\,\text{C} = 7.2 \times 10^{-6}\,\text{N}$

b i $V = \dfrac{Q}{4\pi\varepsilon_0 r} = \dfrac{4.0 \times 10^{-6}\,\text{C}}{4\pi(8.85 \times 10^{-12}\,\text{F m}^{-1}) \times 2.2\,\text{m}} = 1.6 \times 10^4\,\text{V}$

ii $PE = VQ = (1.63 \times 10^4\,\text{V}) \times (2.0 \times 10^{-6}\,\text{C}) = 3.3 \times 10^{-2}\,\text{J}$

Uniform electric fields

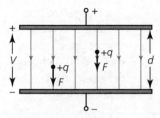

$F = Eq$ is the same at all points in the field

Figure 15

Between a pair of parallel plates with a potential difference applied across them, the electric field is uniform and so the force on a charge anywhere in the field is constant. Suppose that a small charge, δq, is moved from one plate to the other (see Figure 15); then the work done equals force times distance. On the other hand, from the definition of potential difference, the work done in moving the charge is $V \times \delta q$. So:

$$W = F \times d = V \times \delta q$$

Hence:

$$\frac{F}{\delta q} = \frac{V}{d}$$

i.e. $E = \dfrac{V}{d}$

Worked example

a Show that the units $N\,C^{-1}$ and $V\,m^{-1}$ are consistent.

b

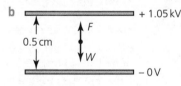

Figure 16

A charged oil drop is held in a fixed position between a pair of parallel plates. The plates are 0.50 cm apart and have a potential difference of 1.05 kV across them (see Figure 16). The oil drop has a radius of 1.2×10^{-6} m and a density of $960\,kg\,m^{-3}$.

  i Determine the field strength between the plates.

  ii Calculate the charge on the oil drop.

Answer

a $V \equiv J\,C^{-1}$ and $J \equiv N\,m$, so $V\,m^{-1} \equiv (N\,m\,C^{-1})m^{-1} \equiv N\,C^{-1}$

b i $E = \dfrac{V}{d} = \dfrac{1.05 \times 10^{3}\,V}{0.50 \times 10^{-2}\,m} = 2.1 \times 10^{5}\,N\,C^{-1}$

Knowledge check 11

Calculate the size of **a** the electric field strength and **b** the electric potential at a distance of 25 cm from a point charge of + 500 pC.

Exam tip

Between the parallel plates the field strength is the same at all points and the force on a charged particle will be the same, wherever it is in the field. It is a common error for students to state that the force on a positive charge increases when it approaches the negative plate.

ii For equilibrium:

electric force = weight of oil drop

$$Eq = mg$$

$$q = \frac{mg}{E} = \frac{\frac{4}{3}\pi(1.2 \times 10^{-6}\,\text{m})(960\,\text{kg m}^{-3}) \times 9.8\,\text{m s}^{-2}}{2.1 \times 10^{5}\,\text{N C}^{-1}} = 3.2 \times 10^{-19}\,\text{C}$$

Capacitance

Capacitors

A capacitor is a device that can store charge. Any isolated conductor can behave as a capacitor, but most commercial capacitors consist of a pair of parallel plates separated by an insulating medium.

The ability of a capacitor to store charge depends on its dimensions, the nature of the insulating material and the potential difference applied across the plates. This is expressed as the **capacitance** C of the capacitor:

$$C = \frac{Q}{V} \qquad \text{unit: farad (F)}$$

where Q is the charge stored when the capacitor is charged to a potential V.

Worked example

Calculate the charge stored on a $220\,\mu\text{F}$ capacitor when the voltage applied is $12\,\text{V}$.

Answer

$$Q = CV = 220 \times 10^{-6}\,\text{F} \times 12\,\text{V} = 2.6 \times 10^{-3}\,\text{C} = 2.6\,\text{mC}$$

Measuring charge

Charge can be measured using a coulomb meter (Figure 17).

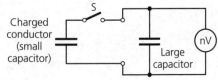

Figure 17

A coulomb meter consists of a capacitor whose known value of capacitance must be much larger than that of the conductor from which the charge is taken. For example, when a charged $1\,\text{F}$ capacitor is connected to an uncharged $100\,\text{F}$ capacitor, the charge will be shared between them, with 100 times more going to the $100\,\text{F}$ capacitor — in effect, virtually all of the charge is transferred to the larger capacitor. By measuring the voltage across the known capacitor the charge can be calculated from $Q = CV$. In practice because the charge, and hence the voltage, is usually extremely small, a DC amplifier is used and the voltmeter is calibrated with the appropriate coulomb scale.

Energy stored in a capacitor

If a charged capacitor is discharged through a lamp or motor, the energy to light the lamp or drive the motor is stored on the capacitor (Figure 18).

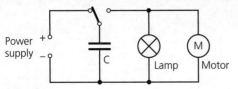

Figure 18

From the $Q = CV$ relationship, a graph of V against Q for a capacitor will be a straight line through the origin. Suppose that a small charge δq is added to the capacitor when it is at voltage V^*; then the work done will be $V^*\delta q$, which is the area of the thin strip of width δq and height V^* shown in Figure 19. The total area beneath the line up to charge Q can be thought of as the sum of the areas of many such thin strips, and this sum of areas is equivalent to the total work done in charging the capacitor up to that point — that is, the energy stored in the capacitor.

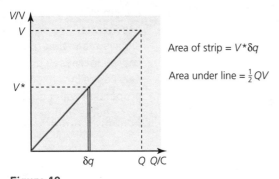

Area of strip = $V^*\delta q$

Area under line = $\frac{1}{2}QV$

Figure 19

From the graph:

E = area under the line up to $(Q,V) = \frac{1}{2}QV$

Using $Q = CV$ to substitute for Q or for V, we find that:

$$E = \tfrac{1}{2}CV^2 = \tfrac{1}{2}\frac{Q^2}{C}$$

Exam tip

Details of the operation of a coulomb meter are not needed for the examination, but questions on the principles of its operation may be set.

Exam tip

The area under any voltage–charge graph represents the energy stored on the capacitor, but calculations will be confined to straight-line graphs.

Worked example

A 2200 µF capacitor is charged to 12 V.

a Calculate the energy stored on the capacitor.

b When connected to a motor with a 20 g mass suspended from a thread wound around the spindle, the capacitor drives the motor raising the mass through 30 cm. Determine the efficiency of the system.

Answer

a $E = \frac{1}{2}CV^2 = \frac{1}{2}(2200 \times 10^{-6}\,\text{F}) \times (12\,\text{V})^2 = 0.16\,\text{J}$

b work done $= mgh = (20 \times 10^{-3}\,\text{kg}) \times 9.8\,\text{m s}^{-2} \times 0.30\,\text{m} = 0.059\,\text{J}$

   efficiency $= \dfrac{0.059\,\text{J}}{0.16\,\text{J}} \times 100 = 37\%$

Charge and discharge of capacitors

When a capacitor is charged or discharged through a *resistor*, the charge (as well as the voltage) rises or falls *exponentially* (see Figure 20).

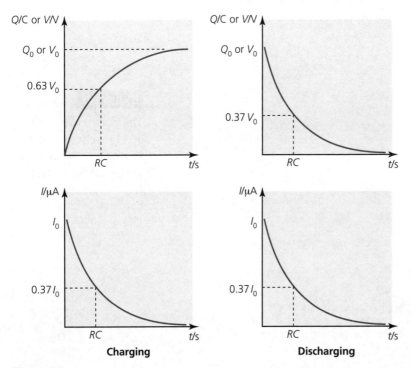

Charging   **Discharging**

Figure 20

Exam tip

You need to be aware that, in this case, an exponential rise or decay means that the charge increases or decreases by a fixed proportion with equal increments of time.

The charge (or voltage) against time graphs (Figure 20) show that the growth or decay of charge is rapid initially, but then reduces as the capacitor becomes charged or discharged.

For a *discharging* capacitor:

$$Q = Q_0\, e^{-t/RC} \quad \text{and} \quad V = V_0\, e^{-t/RC}$$

You will not be required to use the formula for charging a capacitor, $Q = Q_0\,(1 - e^{-t/RC})$.

For both the charge and the discharge of a capacitor, the gradient of the graph of Q against t decreases in magnitude with time. This gradient represents the rate of charge flowing on or off the capacitor, i.e. the current, and we have:

$$I = I_0\, e^{-t/RC} \quad \text{(for both charging and discharging)}$$

From the exponential term $e^{-t/RC}$ you can see that the rate at which a capacitor charges or discharges through a resistor depends on the values of R and C. If the product RC is large, the decay will be slow; RC is known as the **time constant** of the circuit.

When $t = RC$:

$$Q = Q_0\, e^{-RC/RC} = Q_0\, e^{-1} \approx 0.37 Q_0$$

So over a time interval of duration equal to one time constant, the charge (and the voltage and the current) will fall to 37% of the initial value. This fact is the basis of most electronic timing circuits (e.g. the circuit controlling the time delay for a camera) — the voltage across a discharging capacitor falls until a certain switching threshold is reached.

> The **time constant** for a capacitor charging or discharging through a resistor is the product RC.

Core practical 11

Core practical 11 requires you to use an oscilloscope or data logger to display and analyse the potential difference across a capacitor as it charges and discharges through a resistor. You may be asked to provide full details of the circuit used and how the time constant is measured from graphs of potential difference (V) against time (t) or $\ln V$ against t, in the examination.

> **Knowledge check 12**
>
> Show that the unit of the product RC is second.

> **Exam tip**
>
> It is a mathematical requirement that expressions using natural logarithms can be used to determine the exponents of these equations.

Worked example

a Explain why the unit of the time constant for a capacitor discharging through a resistor is the second.

A capacitor is charged to a voltage of 10.0 V and then discharged through a $100\,\mathrm{k\Omega}$ resistor. Measurements of the voltage, taken every 10 s, are shown in Table 1.

Table 1

t/s	0	10	20	30	40	50	60
V/V	10.0	6.3	4.0	2.6	1.6	1.0	0.7

b Plot a graph of voltage against time for the discharging capacitor.

c Use your graph to determine the time constant for the circuit. Hence calculate the value of the capacitance of the capacitor. →

Answer

a t/RC is a numerical power and so does not have a unit. This means that the unit of RC must be that of t, i.e. the second.

b

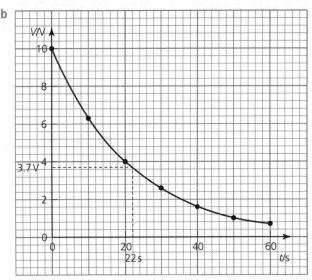

Figure 21

c The voltage will have fallen to 3.7 V (37% of the initial value 10.0 V) after one time constant, which is about 22 s from the graph.

$$\text{So } RC = 22\,\text{s} \Rightarrow C = \frac{22\,\text{s}}{100 \times 10^3\,\Omega} = 2.2 \times 10^{-4}\,\text{F} = 220\,\mu\text{F}$$

Exam tip

An alternative method of finding the time constant is to take the natural logarithm of the voltage and plot $\ln V$ against t. Using the laws of logarithms, $V = Ve^{-t/RC}$ can be written as:

$$\ln V = \ln V_0 - \frac{t}{RC}$$

so the graph of $\ln V$ against t will be a straight line with gradient $-1/RC$.

Summary

After studying this section, you should be able to:

- understand the concept of an electric field, draw the field lines both between charged plates and around a point charge, and use the expression $E = F/Q$
- show that the field strength in the uniform field between charged plates can be found using $E = V/d$
- use Coulomb's law to determine the forces between point charges and to deduce the value of electric field strength at a distance r from the point
- use the equation $V = Q/4\pi\varepsilon_0 r$ for the electric potential at a distance r from a point charge Q

- define *capacitance* and calculate the energy stored in a capacitor for a given potential difference and charge
- sketch curves showing how the charge, potential difference and current vary with time when a capacitor charges and discharges through a resistor; use the curves to determine the *time constant RC* for the circuit
- use the equation $Q = Q_0 e^{-t/RC}$ and similar expressions for potential difference and current to calculate values of t, R and C directly or by using a log–linear graph.

Magnetic fields

We think of a **magnetic field** as a region in which a magnetic force will be experienced by a magnetic material. You are probably familiar with field patterns around bar magnets and coils carrying electric currents. However, unlike electric fields it is less clear what is experiencing the force in a magnetic field. It is actually moving charges.

The moving charges are usually in current-carrying wires, or in beams of electrons or ions.

Magnetic field strength (flux density)

The strength of a magnetic field is defined in terms of the force experienced by a current-carrying wire placed at right angles to the field (Figure 22).

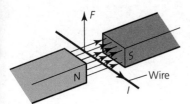

Figure 22

Fleming's left-hand rule (Figure 23) gives the *direction* of the force acting on the wire with reference to the direction of the current and the direction of the field.

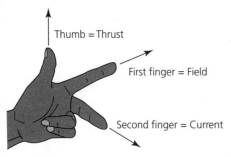

Thumb = Thrust

First finger = Field

Second finger = Current

Figure 23

Experiments show that the *magnitude* of the force acting on the wire is proportional to the current I flowing in the wire and the length l of wire in the magnetic field, i.e. $F \propto Il$, or:

$$F = BIl$$

The constant B reflects the strength of the field and is called the **magnetic flux density** of the field:

$$B = \frac{F}{Il} \quad \text{unit: tesla (T)}$$

If the wire is at an angle θ to the field, then the component of its length at right angles to the field is used to determine the magnitude of the force:

$$F = BIl \sin \theta$$

> A **magnetic field** is a region where a moving charge experiences a force.

> **Exam tip**
>
> Magnetic fields only affect moving charges; electric fields affect both stationary and moving charges.

> **Exam tip**
>
> Moving charges experience a force at right angles to a magnetic field; but in an electric field the force is always in the direction of the field.

> **Knowledge check 13**
>
> Calculate the force acting on 60 cm of wire carrying a current of 1.2 A at right angles to a magnetic field of strength 4.0×10^{-4} T.

Worked example

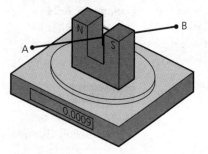
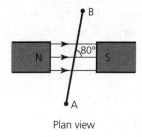

Plan view

Figure 24

In an experiment to determine the force acting on a current-carrying wire in a magnetic field, a magnet is placed on a sensitive balance, and a stiff copper wire is rigidly clamped so that it passes between the poles (Figure 24). The balance is zeroed, and then a current of 2.0 A is passed through the wire. The balance now reads +1.110 g.

a What is the magnitude of the force exerted by the wire on the balance?

b State the direction in which the current is flowing.

c If the wire is at an angle of 80° to the field and has a length of 5.0 cm within the field, calculate the flux density between the poles of the magnet.

Answer

a $F = mg = (1.110 \times 10^{-3}\,\text{kg}) \times 9.81\,\text{m s}^{-2} = 1.1 \times 10^{-2}\,\text{N}$

b The wire exerts a downward force on the magnet, so, by Newton's third law, the magnet exerts an equal upward force on the wire. Using Fleming's left-hand rule, the current must flow from B to A.

c $B = \dfrac{F}{Il \sin \theta} = \dfrac{1.1 \times 10^{-2}\,\text{N}}{2.0\,\text{A} \times 0.050\,\text{m} \times \sin 80°} = 0.11\,\text{T}$

Deflection of charged particles in magnetic fields

Moving charges constitute an electric current, and so will experience a force when they move at a non-zero angle to a magnetic field. Beams of electrons or ions will therefore be deflected as they pass through a magnetic field, as will fast-moving, charged subatomic particles.

The force on a particle carrying a charge q moving with velocity v at an angle θ to a field of flux density B is given by the expression:

$F = Bqv \sin \theta$

Note that because the force is always at right angles to the direction of motion of the charges, particles moving perpendicular to the field will travel along circular paths, with the magnetic force providing the centripetal acceleration. The identification of particles by observing the paths they take in a magnetic field is an important technique in particle physics, which you will look at in more detail later.

A comparison of magnetic fields and electric fields is given in Table 2.

Table 2

Electric field	Acts on both stationary and moving charges	The force acts in the direction of the field
Magnetic field	Acts only on moving charges	The force acts at right angles to the field

Worked example

A beam of electrons is fired between a pair of parallel plates that are 3.0 cm apart in a vacuum, as shown in Figure 25.

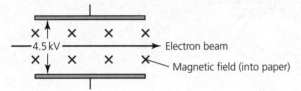

Figure 25

A potential difference of 4.5 kV is applied across the plates, and a uniform magnetic field of flux density 0.20 T is set up between the plates, at right angles to the direction of motion of the electrons.

a Determine the electric field strength between the plates and hence calculate the force acting on an electron in this field.

b Write an expression for the magnetic force on an electron and state the direction of the force if the magnetic field is as shown in Figure 25 (i.e. into the paper).

c Some electrons pass through the fields without being deviated. Explain why this is and state the polarity of the voltage across the plates.

d Calculate the velocity of the undeviated electrons.

Answer

a $E = \dfrac{V}{d} = \dfrac{4.5 \times 10^{-3}\,\text{V}}{3 \times 10^{2}\,\text{m}} = 1.5 \times 10^{5}\,\text{V m}^{-1} = 1.5 \times 10^{5}\,\text{N C}^{-1}$

$F = Eq = 1.5 \times 10^{5}\,\text{N C}^{-1} \times 1.6 \times 10^{-19}\,\text{C} = 2.4 \times 10^{-14}\,\text{N}$

b $F = Bqv$ (the beam is at right angles to the magnetic field). By Fleming's left-hand rule, the force is downward.

c Some electrons are not deviated because the electric and magnetic forces acting on them are equal and opposite. Because the magnetic force is downward, the voltage across the plates must be such that the upper plate is positive in order to generate an upward force on such electrons.

d Balancing the electric and magnetic forces:

$$Eq = Bqv \Rightarrow v = \frac{E}{B} = \frac{1.5 \times 10^{5}\,\text{N C}^{-1}}{0.20\,\text{T}} = 7.5 \times 10^{5}\,\text{m s}^{-1}$$

Exam tip

For a flow of negatively charged electrons, the conventional current (as used in the Fleming rule) is in the opposite direction to the motion of the electrons, i.e. leftward in this case.

Flux and flux linkage

Magnetic fields are represented by 'field lines' or 'lines of force' in the same way as electric fields. The lines are close together where the field is strong, and spread out as the field weakens; this can be seen in Figure 26, which shows the magnetic fields around a bar magnet and a current-carrying solenoid.

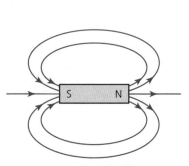

Field is strongest near the poles

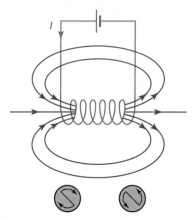

The end where the current flows in a clockwise sense behaves like the south pole of a magnet

Figure 26

The magnetic flux density B can be thought of as the concentration of field lines, or the concentration of the '**magnetic flux**'. Earlier the magnetic flux density was defined in terms of the force on a current-carrying wire, but it can also be thought of as the magnetic flux per unit area:

$$B = \frac{\phi}{A}$$

where ϕ denotes the flux and A is the area of a surface perpendicular to the magnetic field.

Magnetic flux, $\phi = BA$
unit: weber (Wb)

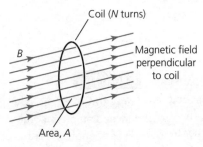

Coil (N turns)

B

Magnetic field perpendicular to coil

Area, A

Figure 27

If a coil with N turns is placed with its edges perpendicular to a magnetic field (Figure 27), the total flux linked with the coil will be $N\phi$. This is called the **flux linkage**.

Flux linkage = $N\phi = NBA$

Electromagnetic induction

Electromagnetic induction is the generation of a current in a conductor by the interaction of a changing magnetic field with the conductor. If a length of conducting wire moves at right angles to a magnetic field, an e.m.f is created across the ends of the wire. This e.m.f exerts a force on the charges in the wire, thereby inducing a current. The direction of the induced current is determined using **Fleming's right-hand rule**.

Electromagnetic induction can also be demonstrated using magnets and coils of wire (see Figure 28).

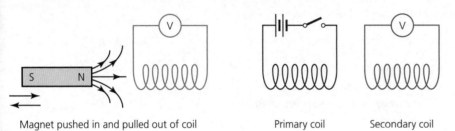

Magnet pushed in and pulled out of coil   Primary coil   Secondary coil

Figure 28

When the magnet is pushed into the coil, an e.m.f. is generated across the coil and this is displayed on the voltmeter. When the magnet is removed, an e.m.f. of the opposite polarity will be observed. It is important to realise that no e.m.f. will be induced while the magnet is stationary inside the coil — there needs to be a *change in flux linkage* for electromagnetic induction to happen. It is helpful to imagine the field lines of the magnet 'cutting through' the wire of the coil while the flux is changing.

> **Exam tip**
>
> There must be some component of the motion of a conductor at right angles to a magnetic field for an e.m.f. to be generated in the conductor.

Now suppose the magnet is replaced by a 'primary' coil connected to a power supply via a switch. When the circuit is switched on, a magnetic field will be formed and the increasing flux will link with the secondary coil generating an e.m.f. across the secondary coil. On switching the circuit off, the magnetic field collapses and the induced e.m.f. in the secondary coil will be of the opposite polarity. Once again, when the current in the primary coil is constant there will be no change in flux linkage and therefore no e.m.f. induced across the secondary coil.

In the magnet–coil experiment, the *magnitude* of the induced *voltage* can be increased by:
- using a stronger magnet
- increasing the rate at which the magnet is inserted or removed from the coil
- increasing the number of turns in the coil.

This can be understood in terms of **Faraday's law** of electromagnetic induction.

The *direction* of flow of the induced *current* is a consequence of the law of conservation of energy. When the north pole of a magnet is pushed into a coil, work is done that transfers electrical energy to the circuit. The current must therefore

Fleming's right-hand rule states that if the direction of a magnetic field is represented by the first finger of the right hand and the motion of the conductor by the thumb, the current direction will be represented by the second finger when the three fingers are mutually at right angles to each other. (See Figure 23 for the left-hand rule diagram.)

Knowledge check 14

A metal rod, held horizontally in an east–west direction, is dropped so that it moves at right angles to the horizontal component of the Earth's magnetic field. Given that the field is in the direction south to north, determine the direction of the induced e.m.f across the ends of the rod.

Faraday's law of electromagnetic induction states that the magnitude of the induced e.m.f. is directly proportional to the rate of change of flux linked with the conductor.

Content Guidance

flow in such a direction that it generates a magnetic field that opposes the entry of the magnet's north pole — the current in the coil should be flowing anticlockwise so that the end where the magnet is entering behaves like another 'north pole' to push against the magnet being inserted. When the magnet is withdrawn from the coil, the induced current creates a 'south pole' (with current flowing clockwise) at the end of the coil to oppose removal of the magnet. This behaviour illustrates **Lenz's law** of electromagnetic induction.

Faraday's and Lenz's laws combine to give a formula for the induced e.m.f. ε:

$$\varepsilon = -\frac{d(N\phi)}{dt} \text{ (the minus sign indicating Lenz's law)}$$

> **Exam tip**
>
> Although this equation is usually written in differential form, all calculations in the examination will give you discrete values of flux linkage and also time to work with. So it is worth remembering the formula as a word equation:
>
> $$\text{e.m.f} = \frac{\text{change in flux linkage}}{\text{time}} = \frac{\Delta(N\phi)}{\Delta t}$$

Worked example

A circular coil of radius 20 cm with 100 turns of wire is placed vertically with its faces along the north–south direction.

a If the horizontal component of the Earth's magnetic field has flux density 2.0×10^{-5} T, calculate the flux linked to the coil.

b The coil is rotated rapidly through 180° about its vertical axis. Determine the change in flux linkage.

c If the time taken for the rotation is 0.06 s, calculate the e.m.f. induced in the coil.

Answer

a flux linkage $= N\phi = NBA$

$$= 100 \times (2.0 \times 10^{-5} \text{ T}) \times \pi \, (20 \times 10^{-2} \text{ m})^2$$

$$= 2.5 \times 10^{-4} \text{ Wb}$$

b After a 180° rotation the faces of the coil will be pointing in opposite directions relative to their original orientation, so the flux will be linked in the reverse direction:

$$N\phi = -2.5 \times 10^{-4} \text{ Wb}$$

Hence the change in flux linkage $= -2.5 \times 10^{-4} \text{ Wb} - 2.5 \times 10^{-4} \text{ Wb}$
$$= -5.0 \times 10^{-4} \text{ Wb}$$

c induced e.m.f. $= \dfrac{\text{change in flux linkage}}{\text{time}} = \dfrac{-5.0 \times 10^{-4} \text{ Wb}}{0.06 \text{ s}} = 8.3 \times 10^{-3} \text{ V}$

Lenz's law of electromagnetic induction states that the current induced in a conductor always flows in such a direction as to oppose the change producing it.

Applications of electromagnetic induction

There are many applications of electromagnetic induction in everyday life. Large rotating coils in magnetic fields are used to generate electricity. Alternating currents in the primary coils of transformers provide a continually changing flux that, when linked with different-sized secondary coils, can step up or step down the voltage. This is needed in the power transmission process, and also for charging laptop computers and mobile phones. In motor cars with petrol engines the spark to ignite the fuel comes from rapidly collapsing a field linked to the ignition coil. This generates a high voltage that produces the spark in the spark plugs.

Lenz's law plays a role in electromagnetic braking, which is used in trains and heavy road vehicles. Electric cars use their motors as generators when slowing down. In addition to the braking effect, this has the advantage of transferring the kinetic energy of the car to charge up the battery.

Alternating current

In Topic 1 (Electric circuits) in core physics you studied the properties of an electric current that flowed in only one direction around a circuit. This flow of charge is described as a **direct current (DC)**.

The current that flows through our household appliances from the mains supply is **alternating current (AC)**.

In a simple generator, an alternating e.m.f. is induced when a coil is rotated continuously in a uniform magnetic field. A graph of the output potential difference against time is shown in Figure 29(a).

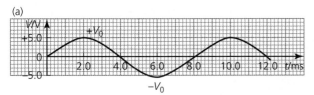

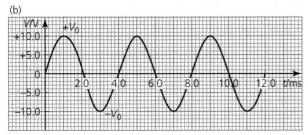

Figure 29

The number of cycles per second equals the frequency of rotation of the coil in the field, and the maximum values of the potential difference ($\pm V_0$) are called the **peak values**. Figure 29(b) shows the output potential difference for the situation when the frequency of rotation is twice that in Figure 29(a). The rate of change of flux linkage has doubled, so the peak values have doubled, the **frequency** of the alternating potential difference has doubled and the **period** is halved.

A **direct current** is a flow of charge that moves in one direction around a circuit.

An **alternating current** occurs when the direction of the current changes periodically.

The **frequency** of an alternating potential difference (current) is the number of complete cycles per second.

The **period** of an alternating potential difference (current) is the time for one complete cycle.

The **peak value** of an alternating potential difference (or current) is the maximum value in one direction or the other.

For a coil rotating with a frequency, f, the value of the potential difference, V, at an instant is given by:

$$V = V_0 \sin 2\pi f t$$

If the output of the generator is connected across a resistor, an alternating current will flow through the resistor. Its value is given by:

$$I = I_0 \sin 2\pi f t$$

An alternating current (AC) flowing in a resistor will display all the properties of a direct current. As its value is continuously changing we need to assign an effective value to the AC, i.e. the direct current (DC) that would have the same effect as the AC. This can be achieved by comparing the average power dissipated by the AC with that of a steady DC.

It can be shown that the effective value of an AC is equal to its **root mean square value**, I_{rms}.

■ Power dissipated by a steady DC, $P = I^2 R$
■ Power dissipated by an AC, $P = I_{rms}^2 R$

> The **root mean square value** of an alternating current or potential difference equals the peak value divided by the square root of 2:
>
> $$I_{rms} = \frac{I_0}{\sqrt{2}}; \qquad V_{rms} = \frac{V_0}{\sqrt{2}}$$

Knowledge check 15

Calculate **a** the peak voltage of a $230\,V_{rms}$ mains supply and **b** the rms value of an alternating current with a peak value of $5.0\,A$.

Worked example

For the alternating potential differences in Figures 29(a) and 29(b) calculate:

a the period of the alternating potential difference
b the frequency of the alternating potential difference.

The output is connected across a $50\,\Omega$ resistor. For each output calculate:

c the peak value of the current
d the rms value of the current
e the power dissipated in the resistor.

Answer

a $T = 8.0\,ms$ for Figure 29(a); $T = 4.0\,ms$ for Figure 29(b)

b $f = \dfrac{1}{T} = \dfrac{1}{8 \times 10^{-3}\,s} = 125\,Hz$ for Figure 29(a); $f = 250\,Hz$ for Figure 29(b)

c $I_0 = \dfrac{V_0}{R} = \dfrac{5.0\,V}{50\,\Omega} = 0.10\,A$ for Figure 29(a); $\dfrac{V_0}{R} = \dfrac{10\,V}{50\,\Omega} = 0.20\,A$ for Figure 29(b)

d $I_{rms} = \dfrac{I_0}{\sqrt{2}} = \dfrac{0.10\,A}{\sqrt{2}} = 0.071\,A$ for Figure 29(a); $I_{rms} = \dfrac{0.20\,A}{\sqrt{2}} = 0.14\,A$ for Figure 29(b)

e $P = I_{rms}^2 R = (0.071\,A)^2 \times 50\,\Omega = 0.25\,W$ for Figure 29(a); $P = (0.14\,A)^2 \times 50\,\Omega = 0.98\,W$ for Figure 29(b) or $1.0\,W$ if data kept in calculator.

Summary

After studying this section, you should be able to:

- understand the concept of a magnetic field and be able to draw the field lines for a magnet, a coil and a solenoid
- use the expressions $F = BIl\sin\theta$ for a current-carrying wire in a magnetic field and $F = Bqv\sin\theta$ for a charged particle moving in a field; apply Fleming's left-hand rule to determine the relative directions of force, field and current
- define flux linkage and explain how an e.m.f. is induced in a conductor when it moves relative to a

magnetic field; apply Fleming's right-hand rule to determine the relative directions of field, motion and induced current
- define and use Faraday's and Lenz's laws of electromagnetic induction
- understand what is meant by the terms *frequency*, *period*, *peak value* and *root mean square value* when applied to alternating currents and alternating potential differences
- be able to use the equations $V_{rms} = V_0/\sqrt{2}$ and $I_{rms} = I_0/\sqrt{2}$.

Nuclear and particle physics

This topic covers atomic structure, particle accelerators and the quark–lepton model of fundamental particles.

The nuclear atom

You will have learnt about the Rutherford model of an atom, which has the bulk of the matter contained in a relatively small nucleus consisting of protons and neutrons, with electrons orbiting around it.

Each element is identified by its **proton number** and each isotope by the total number of protons and neutrons — the **nucleon number**. Consider this equation:

$$^{238}_{92}U \rightarrow {}^{234}_{90}Th + {}^{4}_{2}He$$

Uranium has proton number 92 and nucleon number 238, so it has 92 protons and 146 neutrons. On emission of an alpha particle, which has two protons and two neutrons, the proton number falls by two and the nucleon number by four, leaving thorium-234.

In all nuclear transformations, the total proton and nucleon numbers on both sides of the equation must balance.

Evidence for the nuclear atom

The evidence for Rutherford's nuclear model of an atom was provided by the large-angle alpha particle scattering experiment performed by Geiger and Marsden (Figure 30). Alpha particles were fired at a thin sheet of gold. Most of the particles passed through the film without being deviated but a small number were scattered, including some that were deviated through large angles (bigger than 90°).

Knowledge check 16

Give the number of protons and the number of neutrons in the following nuclei: $^{14}_{6}C$, $^{235}_{92}U$.

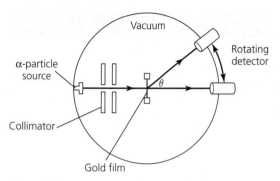

Figure 30

In the examination, you may be asked to give full details of the experimental set-up; this information can be found in most textbooks. Below is an outline of the principles and major conclusions.

- Alpha particles are about 7500 times as massive as electrons, so their paths will be little affected by collisions with electrons.
- The diameter of a nucleus is very much less than the diameter of its atom; so, in a thin sheet of gold the probability of electrons passing close to the nuclei is quite low —atoms are mostly empty space.
- Alpha particles are positively charged ($+2e$, where e is the magnitude of charge on an electron), and so are the gold nuclei ($+79e$). So the force between an alpha particle and a gold nucleus is repelling and will obey an inverse square law (Coulomb's law), decreasing rapidly with distance between the charged particles.
- If an alpha particle makes a direct collision with a nucleus, it can be scattered back towards the source.
- Detailed analysis of the numbers of particles scattered through a range of angles supports the model of a small, positive nucleus surrounded by a much larger envelope of orbiting electrons.

Worked example

An alpha particle of mass 6.7×10^{-27} kg moving at $1.0 \times 10^7\,\text{m s}^{-1}$ is fired directly at the nucleus of a gold atom.

a Calculate the average force needed to bring the alpha particle to rest, if it is stopped in a distance equal to the radius of the gold atom (1.3×10^{-10} m).

b Use Coulomb's law to determine the mean distance from the gold nucleus for such a force to act on the alpha particle.

c Comment on how your answer supports the experimental evidence for the nuclear atom.

Answer

a Using $v^2 = u^2 + 2as$, we have:

$$a = \frac{v^2 - u^2}{2s} = \frac{(0\,\text{m s}^{-1})^2 - (1.0 \times 10^7\,\text{m s}^{-1})^2}{2(1.3 \times 10^{-10}\,\text{m})} = -3.8 \times 10^{23}\,\text{m s}^{-2}$$

Hence $F = ma = (6.7 \times 10^{-27}\,\text{kg}) \times (3.8 \times 10^{23}\,\text{m s}^{-2}) = 2.6 \times 10^{-3}\,\text{N}$ ➔

Exam tip

There are three main observations from this experiment. To gain 3 marks in an examination the italicised words below must be included.

Most of the alpha particles pass *straight through*; *some* alpha particles are deflected; some, *very few*, are deviated through large angles (or straight back).

Exam tip

The three deductions that can be made using these observations are:

- *most* of the atom is *empty* space
- most of the mass of the atom is concentrated in a *tiny* nucleus
- the nucleus is (positively) *charged*.

Just saying that the nucleus is positive will lose a mark.

b From Coulomb's law:

$$r^2 = \frac{Q_1 Q_2}{4\pi\varepsilon_0 F} = \frac{(2 \times 1.6 \times 10^{-19}\,\text{C}) \times (79 \times 1.6 \times 10^{-19}\,\text{C})}{(4\pi) \times (8.85 \times 10^{-12}\,\text{F m}^{-1}) \times (2.6 \times 10^{-3}\,\text{N})}$$

and this gives $r = 3.7 \times 10^{-12}\,\text{m}$

c Because the force that a gold nucleus exerts on an alpha particle falls sharply with distance according to an inverse square law, for the alpha particle to be stopped it needs to pass extremely close to the gold nucleus. The distance found in part (b) is very small (about 3%) compared with the radius of the gold atom. This means that the nucleus must occupy a tiny region relative to the size of the atom, supporting the model of a nuclear atom.

Exam tip

Note that sections of the core physics and the further mechanics topic are often included in particle physics questions.

Development of atomic models

The Rutherford model of the atom was an important step in the development of our understanding of the atom, but the idea of the atom as a fundamental particle was first suggested over 2000 years ago by the Greek philosopher Democritus. In the nineteenth century John Dalton explained some chemical reactions by using the idea that each element consisted of indivisible 'balls' and, after his discovery of the electron, J. J. Thomson proposed a 'plum-pudding' model in which the electrons (corpuscles) were evenly distributed in a 'sea' of positive charge. Rutherford's model superseded all this and later discoveries of the neutron and proton led to the idea that matter consisted of three fundamental particles. The model was further adapted to fit the ideas of quantum theory and wave–particle duality by the likes of Niels Bohr and Erwin Schrödinger during the early part of the twentieth century.

The Rutherford model is still useful for explaining many physical properties and chemical properties, but modern theories and high-energy subatomic collisions have indicated that there are many more fundamental particles. The quark–lepton model is discussed later in this section.

Particle accelerators

Charged particles can be accelerated by electric fields and magnetic fields. One of the simplest particle accelerators is the **electron gun** (Figure 31), which is used in cathode-ray tubes and X-ray tubes. Electrons are produced from a heated metal filament or a metallic oxide surface by thermionic emission; then they are accelerated by an electric field.

Electron gun

- A hot cathode emits electrons by thermionic emission (i.e. electrons near the surface gain sufficient thermal energy to overcome the work function; see the particle properties of light in the second guide in this series covering Topics 4 and 5).
- The electrons experience a force in the electric field between the cathode and the anode, and are accelerated towards the anode:

$F = Eq = ma$ (where q is the electron charge, $1.6 \times 10^{-19}\,\text{C}$)

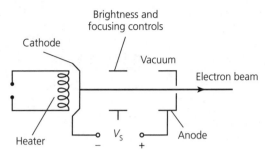

Figure 31

- The work done on the electron by the field equals the gain in kinetic energy:

$$Vq = \frac{1}{2}mv^2$$

- Some of the fast-moving electrons pass through the hole in the anode and strike a screen placed beyond.
- In practice, the gun may have several anodes that can have their voltages adjusted so as to focus the electron beam to a particular spot on the screen.

Worked example

a Show that the speed of an electron leaving an electron gun that has a potential difference of 10 kV between its cathode and the accelerating anode is about $6 \times 10^7\,\mathrm{m\,s^{-1}}$. The mass of an electron is $9.1 \times 10^{-31}\,\mathrm{kg}$.

b Some electrons may leave the gun with speeds higher than the above. Suggest how this is possible.

Answer

a From $Vq = \frac{1}{2}mv^2$ we have:

$$(10 \times 10^3\,\mathrm{V}) \times (1.6 \times 10^{-19}\,\mathrm{C}) = \frac{1}{2}(9.1 \times 10^{-31}\,\mathrm{kg}) \times v^2$$

So $v = 5.9 \times 10^7\,\mathrm{m\,s^{-1}}$

b Some electrons will be emitted with 'extra' kinetic energy transferred from the process of thermionic emission.

Linear accelerator

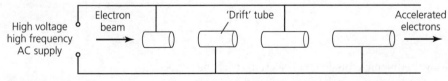

Figure 32

A linear accelerator (linac) operates on the same principle as an electron gun — electrons or other charged particles are accelerated across gaps between charged electrodes (see Figure 32). In a linac there can be as many as 100 000 'drift tubes' connected to a high-frequency, high-voltage AC supply. These are arranged in such

---

Exam tip

The conservation of energy applies in all accelerators. For the electron gun, the kinetic energy gained by an electron $(\frac{1}{2}mv^2)$ equals the work done on the electron by the potential difference between the electrodes (Vq).

Knowledge check 17

Calculate the speed of a doubly charged ion of mass $6.7 \times 10^{-27}\,\mathrm{kg}$ when it has been accelerated from rest across a potential difference of 10 kV.

a way that the particles gain kinetic energy between the tubes and move at constant speed inside the tubes. The main principles are:

■ Alternate tubes are connected to each terminal of the AC supply.
■ The charged particles spend one half of each period of the alternating voltage between two tubes and the other half of each cycle inside one of the tubes.
■ During a half-cycle, when the voltage would oppose their motion, the particles are inside a tube where they are shielded from the electric field; so the particles travel at constant speed within the tube (i.e. they 'drift' through the tube).
■ The particles gain kinetic energy as they travel across successive gaps and can be accelerated to energies as high as 30 GeV.
■ The length of the drift tubes increases along the accelerator so that although the speed of the particles is increasing, the time needed to pass through each tube will always be the same (equal to the half of the period of the alternating electric field).

Cyclotron

The main disadvantage of linear accelerators is that they need to be very long in order to produce high energies. The Stanford linear accelerator in the USA has a length of about 3 km. Cyclotrons, while based on the same principle of synchronous acceleration as linacs, use a *magnetic* field to make the charged particles move in a spiral path.

To understand the action of a cyclotron (Figure 33), it might be helpful to look back at the section on the deflection of charged particles in a magnetic field.

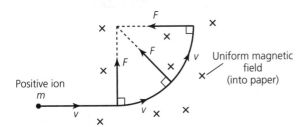

Figure 33

Using Fleming's left-hand rule, a particle carrying charge q moving with speed v at right angles to a magnetic field of flux density B will experience a force of magnitude Bqv in a direction perpendicular to its motion. The particle will therefore follow a circular path and:

$$Bqv = F = \frac{mv^2}{r}$$

so: $r = \dfrac{mv}{Bq} = \dfrac{p}{Bq}$

This relationship ($r = p/Bq$) between the momentum of a charged particle in a magnetic field and the radius of the circular motion also plays an important role in the identification of particles in detectors (see later).

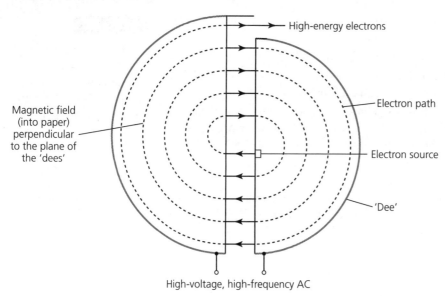

Figure 34

A cyclotron consists of two hollow, semi-circular D-shaped sections ('dees') fixed at right angles to a uniform magnetic field and have a high-frequency alternating voltage applied between them (see Figure 34).

An ion source fires charged particles into the gap between the dees close to the centre.

- The particles are accelerated across the gap during a half-cycle of the alternating voltage when the polarity is appropriate.
- During the next half-cycle, the particles follow a circular path at constant speed in one of the dees.
- The particles are then accelerated across the gap into the other dee.
- Faster particles follow paths of larger radius, so that all particles always spend the same amount of time in the dees and their acceleration is synchronised with the appropriate half-cycle of alternating voltage.

Worked example

A cyclotron has a maximum radius of 0.50 m and uses a magnetic field of strength 1.2 T. It is used to accelerate protons.

a Show that the momentum of the protons as they leave the cyclotron is about $1 \times 10^{-19}\,\text{kg m s}^{-1}$.

b Calculate the kinetic energy of the emitted protons. Give your answer in MeV.

Answer

a $p = Bqr = 1.2\,\text{T} \times (1.6 \times 10^{-19}\,\text{C}) \times 0.50\,\text{m} = 0.96 \times 10^{-19}\,\text{kg m s}^{-1} \approx 1 \times 10^{-19}\,\text{kg m s}^{-1}$

b $E_k = \dfrac{p^2}{2m} = \dfrac{(0.96 \times 10^{-19}\,\text{kg m s}^{-1})^2}{2 \times (1.67 \times 10^{-27}\,\text{kg})} = 2.8 \times 10^{-12}\,\text{J}$

$= \dfrac{2.8 \times 10^{-12}\,\text{J}}{1.6 \times 10^{-19}\,\text{J eV}^{-1}} = 1.7 \times 10^7\,\text{eV or 17 MeV}$

Relativistic effects

For speeds approaching the speed of light, relativistic effects need to be taken into account. For example, suppose that an electron is accelerated across a potential difference of 1 GV; if the electron has a mass of 9.11×10^{-31} kg (the electron mass listed in the data sheet), the equation $Vq = \frac{1}{2}mv^2$ predicts that v would be about 2×10^{10} m s^{-1}. However, it is a basic postulate of the theory of relativity that nothing can travel faster than the speed of light (3×10^8 m s^{-1}). Therefore, for energy to be conserved, the *mass* of the electron must *increase*. The equivalence of mass and energy is discussed later, but for now you should just be aware that relativistic effects can create synchronisation problems in high-energy particle accelerators. **Synchrotrons**, such as CERN in Switzerland, account for these relativistic effects and can produce particles with extremely high energy.

Exam tip

Particle accelerators are needed to provide sufficient energy for particles to overcome electric fields in order to collide with other particles, and to enable the creation of new particles.

Particle detectors

Particles can be detected when they interact with matter to cause ionisation or when they excite electrons to higher energy levels, accompanied by the emission of photons. Although specific details of particle detectors are not required for the examination, you are expected to know the basic principles and applications of ionisation chambers and semiconducting devices.

In bubble tanks and cloud chambers, a charged particle passing through will generate a trail of ions along which bubbles or vapour droplets are formed, making these paths visible. The nature of the particles can be deduced from the length of the trails they leave, and from how these paths are affected by electric fields and magnetic fields.

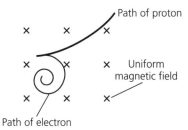

Figure 35

Figure 35 shows a bubble-tank image of the trails produced when a high-energy photon travelling from left to right strikes a neutron, which splits into a proton and an electron. The following information can be gained from the tracks:

- The photon is uncharged, so it does not produce a trail of ions and cannot be observed.
- The proton is positively charged, and Fleming's left-hand rule indicates that the force acting on the proton is initially upward. Because the charge on the electron is negative, it will be deviated in the opposite direction.
- From the relationship $r = p/Bq$ we can infer that the proton has bigger momentum than the electron because its path has a larger radius. Note that a thicker trail also indicates more intense ionisation, which is usually due to the particle having higher momentum or charge.

- The path of the electron spirals inward. This is because the electron loses energy and its momentum decreases and the radius of its path becomes smaller. Although less noticeable from the traces in the diagram, the radius of the proton's track will also decrease as its momentum falls.

Examples of excitation detectors are fluorescent screens, scintillation counters and solid-state (electronic) devices.

Particle interactions

In all interactions, **energy**, **charge** and **momentum** must be conserved. In the bubble-tank example above, the particle interaction can be expressed as a simple equation:

$$n^0 + \gamma \rightarrow p^+ + e^-$$

It is clear that because both the neutron and the photon are uncharged, the initial charge is zero. After the interaction the total charge is also zero because the proton carries a charge of $+e$ (where $e = 1.6 \times 10^{-19}\,C$ — the magnitude of electron charge) and the electron carries a charge of $-e$.

The photon has momentum, so if the neutron was stationary before the interaction, the total momentum of the proton and the electron must be the same as the initial momentum of the photon.

> **Exam tip**
>
> If a single track in a bubble chamber is seen to suddenly change direction, with no other trail observed, it is likely that an uncharged particle or a neutrino has been produced. Because momentum is a vector quantity, a change in direction represents a change in momentum and so the uncharged (unobserved) particle must travel in a different direction so that momentum is conserved.

The energy conservation in this scenario is less clear. The photon has energy hf and the proton and electron will gain kinetic energy after the collision. However, the total mass of the proton and the electron differs from the mass of the neutron. Here, mass–energy equivalence must be accounted for.

The **rest mass** of the particles can be represented as an equivalent **energy** using Einstein's famous equation $E = mc^2$ in the form:

$$m_0 = \frac{E_0}{c^2}$$

The rest mass can be thought of as the energy that would be transferred if the entire mass were to be dematerialised, or if the particle were to be made up completely from other forms of energy.

It is often convenient to represent the rest mass of subatomic particles in terms of the non-SI units MeV/c^2 or GeV/c^2.

> **Exam tip**
>
> Most detectors rely on ionisation to show the tracks of moving particles. Uncharged particles, photons and neutrinos produce little or no ionisation and so are extremely difficult to detect.

> **Exam tip**
>
> It is insufficient to state that the charge is the same before and after an interaction. For charge to be conserved, the magnitude *and* the nature of the charges for every particle should be given. For the example here, $0 + 0 \rightarrow +1 + (-1)$.

> **Knowledge check 19**
>
> Calculate the rest mass energy of an electron.

> **Exam tip**
>
> The mass of moving particles can be taken to be the same as their rest mass, unless the speed is close to the speed of light.

■ One electronvolt (eV) is the work done in moving an electron through a potential difference of one volt:

$1 \, \text{eV} = (1.6 \times 10^{-19} \, \text{C}) \times 1 \, \text{V} = 1.6 \times 10^{-19} \, \text{J}$

$1 \, \text{MeV} = 1.6 \times 10^{-13} \, \text{J}$

$1 \, \text{GeV} = 1.6 \times 10^{-10} \, \text{J}$

MeV and GeV are units of energy; MeV/c^2 and GeV/c^2 are the corresponding units of mass.

Knowledge check 20

Convert a mass of $1.2 \times 10^{-29} \, \text{kg}$ to MeV/c^2.

Worked example

Express the rest mass of

a an electron

b a proton

in terms of MeV/c^2.

Answer

a The rest-mass energy of an electron is given by:

$E_0 = m_0 c^2 = 9.11 \times 10^{-31} \, \text{kg} \times (3.00 \times 10^8 \, \text{m s}^{-1})^2 = 8.20 \times 10^{-14} \, \text{J}$

$= \dfrac{8.20 \times 10^{-14} \, \text{J}}{1.60 \times 10^{-19} \, \text{J eV}^{-1}} = 5.12 \times 10^5 \, \text{eV} = 0.512 \, \text{MeV}$

So the rest mass of an electron is:

$m_0 = 0.512 \, \text{MeV}/c^2$

b The rest-mass energy of a proton is:

$E_0 = m_0 c^2 = 1.67 \times 10^{-27} \, \text{kg} \times (3.00 \times 10^8 \, \text{m s}^{-1})^2 = 1.50 \times 10^{-10} \, \text{J}$

$= \dfrac{1.50 \times 10^{-10} \, \text{J}}{1.60 \times 10^{-19} \, \text{J eV}^{-1}} = 9.39 \times 10^8 \, \text{eV} = 939 \, \text{MeV}$

So the rest mass of a proton is:

$m_0 = 939 \, \text{MeV}/c^2$

Another unit used in particle physics is the **unified atomic mass unit**, denoted by 'u'. It is defined as one-twelfth of the mass of a carbon-12 atom, so:

$1 \, \text{u} = 1.66 \times 10^{-27} \, \text{kg}$

The energy equivalent ($m_0 c^2$) is $1.49(4) \times 10^{-10} \, \text{J}$ or $931 \, \text{MeV}$.

You will often see the concept of mass–energy equivalence expressed as:

$\Delta E = c^2 \Delta m$

where 'Δ' means 'change in'. This is the form given in the formulae sheet.

Knowledge check 21

Calculate the energy equivalent of a particle of mass $4.04 \, \text{u}$.

Creation and annihilation of matter and antimatter

Early work with particle detectors showed that cosmic rays could produce some tracks identical to those of an electron, but which curve in the opposite direction. This was the first piece of evidence for the existence of **antimatter**.

When a photon interacts with the strong electric field around a nucleus, it is transformed into two particles. The underlying process of **pair production** is:

photon → particle + antiparticle

An example of an **antiparticle** is the antielectron, or **positron**, whose mass is identical to that of the electron but it carries the opposite charge (of equal magnitude).

Pair production in this case can be written as:

$\gamma \to e^- + e^+$

As usual, it is essential for momentum to be conserved in the creation of the electron–positron pair. Production of the positron ensures that the net charge remains zero. It is also essential that the gamma-ray photon has at least as much energy as the combined rest-mass energy of the electron and the positron.

Antimatter can be created from collisions in particle accelerators, but very little exists because of the vast amount of energy needed to create it.

> For a charged particle its **antiparticle** has the same mass but opposite charge.

> **Exam tip**
>
> Uncharged particles like neutrons and neutrinos have antiparticles that differ from the particle in more subtle ways, like quark composition or lepton number.

> **Exam tip**
>
> To create a new particle, the energy transferred must be at least equal to the rest-mass energy of the particle. If more energy is transferred the new particle must gain some kinetic energy.

Worked example

A gamma-ray photon interacts with a nucleus to form an electron–positron pair.

a Calculate the minimum energy of the gamma ray required for this interaction.

b If the gamma ray has a wavelength of 1.05×10^{-12} m, calculate the maximum kinetic energy of the electron and positron.

Answer

a A positron and an electron have the same rest-mass energy, which was found to be 0.512 MeV in the previous Worked example. So the minimum energy that the gamma ray needs to have is:

$2 \times 0.512\,\text{MeV} = 1.02\,\text{MeV}$

b Energy of photon $= \dfrac{hc}{\lambda} = \dfrac{(6.63 \times 10^{-34}\,\text{J s}) \times (3.00 \times 10^{-8}\,\text{m s}^{-1})}{1.05 \times 10^{-12}\,\text{m}}$

$= 1.89 \times 10^{-13}\,\text{J} = 1.18\,\text{MeV}$

The maximum kinetic energy of electron and positron $= (1.18\,\text{MeV} + (-\,1.02\,\text{MeV})$

$= 0.16\,\text{MeV (or 0.08 MeV per particle)}$

When matter collides with antimatter, the reverse of the pair-production equation happens:

particle + antiparticle → photon + photon

Antiparticles have a very short lifetime. This is because matter is much more prevalent than antimatter, so antiparticles exist for only a brief time before they meet a particle and are **annihilated** to generate gamma-ray photons.

> **Exam tip**
>
> Two photons must be created by particle–antiparticle annihilation in order to conserve momentum.

The quark–lepton model

Subatomic particles

A subatomic particle is any particle that is smaller than an atom. **Fundamental particles**, also known as **elementary particles**, are subatomic particles that are thought to be indecomposable and which form the building blocks for other subatomic particles. There are two kinds of fundamental particles that you need to know about — **leptons** and **quarks**.

Leptons consist of the electron e^-, the muon μ^- and the tau particle τ^- (each carrying charge $-e$), the electron neutrino ν_e, the muon neutrino ν_μ and the tau neutrino ν_τ (each of charge zero) together with their antiparticles.

Quarks consist of d (down), u (up), s (strange), c (charm), b (bottom) and t (top) particles —their antiparticles are called 'antiquarks'. The d, s and b particles carry charge $-\frac{1}{3}e$, while u, c and t carry $+\frac{2}{3}$.

Before the top quark was discovered, its existence was predicted from the symmetry of the quark–lepton model.

Hadrons are subatomic particles composed of quarks. They are classified into baryons and mesons.

A **baryon** is made up of three quarks. Baryons other than protons and neutrons are very short-lived. Protons carry charge $+e$ and are composed of the quarks uud; neutrons have charge zero and quark composition udd.

A **meson** is made up of a quark and an antiquark. For example, a π^+ meson is made from u and $\bar{\text{d}}$ (anti-d), and a π^- meson from d and $\bar{\text{u}}$ (anti-u). A K^+ meson is made from u and $\bar{\text{s}}$, and a K^- meson from s and $\bar{\text{u}}$.

Photons play an important role in many subatomic interactions and may also be classified as fundamental particles.

Particle interactions and the conservation laws

In high-energy accelerators, such as CERN, there are many interactions involving baryons, mesons and leptons. In all of these, as with large-scale collisions, the laws of conservation of energy and momentum must be obeyed. In addition there are other conservation laws that must be upheld. You will be required to use the laws of conservation of charge Q, baryon number B and lepton number L to check if a given reaction is possible.

Baryons and mesons can have a positive charge $(+1e)$, negative charge $(-1e)$ or zero charge. Electrons, muons and tau particles have negative charges $(-1e)$ (their antiparticles are positively charged) while neutrinos have zero charge.

For a reaction to be possible, the total charge before must be the same as the total charge after the event.

For example, $K^+ + p \rightarrow n + e^-$ is not possible because:

Q: $+1e +(+1e) \neq 0 + (-1e)$

Exam tip

The Edexcel specification does not require you to remember the exact composition of the various hadrons, but you are expected to be familiar with the notation and to be able to interpret equations involving standard particle symbols such as π^+ and e^-.

Knowledge check 22

State the two possible quark structures for a K^0 meson.

However, $\Lambda^0 \to \pi^- + p$ is possible because:

$\quad$ Q: $0 = -1e + (+1e)$

All baryons have a baryon number of 1 (their antiparticles have a baryon number of −1). Mesons and leptons have baryon number 0. For a reaction to be possible, the total of the baryon numbers before the event must equal the total after. For example, $\pi^+ + \pi^- \to n + K^0$ is not possible because:

$\quad$ B: $0 + 0 \neq 1 + 0$ (K and π are mesons and the neutron is a baryon)

But $K^- + p \to K^+ + K^- + \Omega^0$ is possible because:

$\quad$ B: $0 + 1 = 0 + 0 + 1$ (the K particles are mesons and the proton and Ω^0 are baryons)

Leptons are assigned the following numbers:
- electron and electron neutrino $L_e = 1$, $L_\mu = 0$, $L_\tau = 0$
- muon and muon neutrino $L_e = 0$, $L_\mu = 1$, $L_\tau = 0$
- tau and tau neutrino $L_e = 0$, $L_\mu = 0$, $L_\tau = 1$.

All the antiparticles have the appropriate $L = -1$.

For the decay: $\mu^- \to e^- + \bar{\nu}_\mu + \nu_e$

$\quad$ L_e: $0 = 1 + 0 + (-1)$; L_μ: $1 = 0 + 1 + 0$; L_τ: $0 = 0 + 0 + 0$

So all the lepton numbers are conserved and the decay is possible.

Worked example

Use the conservation laws for charge (Q), baryon number (B) and lepton number (L) to determine which of the following reactions cannot occur:

a $\pi^- + p \to n + \pi^0$

b $K^+ + K^- \to \pi^0$

c $n \to p + e^{-1} + \nu_e$

Note that protons, p, neutrons, n, lambda, Λ, epsilon, Σ, and omega, Ω, particles are baryons. Pions, π, and kaons, K, are mesons.

Answer

a $\quad$ Q $\quad$ $-1 + (+1) \to 0 + 0$ $\quad$ $0 = 0$ $\quad$ possible
$\quad\quad$ B $\quad$ $0 + 1 \to 1 + 0$ $\quad\quad$ $1 = 1$ $\quad$ possible
$\quad\quad$ L $\quad$ $0 \to 0$ $\quad$ $0 = 0$ $\quad$ possible

b $\quad$ Q $\quad$ $+1 + (-1) \to 0$ $\quad$ $0 = 0$ $\quad$ possible
$\quad\quad$ B $\quad$ $0 + 0 \to 0$ $\quad$ $0 = 0$ $\quad$ possible
$\quad\quad$ L $\quad$ $0 \to 0$ $\quad$ $0 = 0$ $\quad$ possible

c $\quad$ Q $\quad$ $0 \to (+1) + (-1) + 0$ $\quad$ $0 = 0$ $\quad$ possible
$\quad\quad$ B $\quad$ $1 \to 1 + 0 + 0$ $\quad$ $1 = 1$ $\quad$ possible
$\quad\quad$ L_e $\quad$ $0 \to 0 + 1 + 1$ $\quad$ $0 \neq 2$ $\quad$ not possible

Reaction (c) is not possible because the lepton numbers are not conserved. (This is β^- decay where the beta particle is accompanied by an antineutrino. This will be covered in the guide covering Topics 9–13 in this series.)

Summary

After studying this section you should be able to:

- describe the structure of the nuclear atom, and explain how alpha particle deflection experiments support this model
- explain how particles are accelerated in linear devices and cyclotrons, and use the equations $Vq = \frac{1}{2}mv^2$ and $r = p/Bq$ to determine the speeds of particles
- give details of particle detectors and interpret the tracks for given interactions
- understand the meanings of standard particle symbols and apply the conservation laws to a range of subatomic particle interactions
- recognise a list of fundamental particles and their antiparticles; state the difference in quark structure between baryons and mesons; understand that all leptons are accompanied by neutrino or antineutrino formations

Questions & Answers

The Edexcel examinations

The Edexcel A-level physics examination consists of three papers. Paper 1 and paper 2 contain multiple-choice, short-open, open-response, calculations and extended-writing questions. Both papers are of 1 hour and 45 minutes duration and have 90 marks.

Paper 1 covers all the topics in core physics I (Mechanics and electric circuits) and those in advanced physics I (Further mechanics, Electric and magnetic fields, and Nuclear and particle physics).

Paper 2 covers the sections in core physics II (Materials and Waves and the particle nature of light) and those in advanced physics II (Thermodynamics, Space, Nuclear radiation, Gravitational fields and Oscillations).

Paper 3 covers the general and practical principles of physics. It is of 2 hours and 30 minutes duration and has 120 marks. This paper covers all of the topics and includes synoptic questions as well as assessing the conceptual and theoretical understanding of experimental methods.

All papers will also examine 'Working as a physicist'. Briefly this means students:
- working scientifically, developing competence in manipulating quantities and their units, including making estimates
- experiencing a wide variety of practical work, developing practical and investigative skills by planning, carrying out and evaluating experiments and knowing about the ways in which scientific ideas are used
- developing the ability to communicate their knowledge and understanding of physics
- acquiring these skills through examples and applications from the entire course

This guide covers only the sections on Further mechanics, Electric and magnetic fields, and Nuclear and particle physics that are required for the A-level paper 1 and paper 3.

A formulae sheet is provided with each test. Copies can be downloaded from the Edexcel website, or can be found at the end of past papers.

Command terms

Examiners use certain words that require you to respond in a particular way. You must be able to distinguish between these terms and understand exactly what each requires you to do. Some frequently used commands are shown below (a full list can be found in the Edexcel specification).
- **State** — a brief sentence giving the essential facts; no explanation is required (nor should you give one).
- **Define** — you can use *word equations*; if you use *symbols* you must state what each represents.

- **List** — write a series of words or terms, with no need to write sentences.
- **Outline** — a logical series of bullet points or phrases will suffice.
- **Describe** — for an experiment a diagram is essential, then give the main points concisely (bullet points can be used).
- **Draw** — diagrams should be drawn in section, neatly and *fully labelled* with all measurements clearly shown, but don't waste time — remember it is not an art exam.
- **Sketch** — usually a graph, but graph paper is not necessary, although a grid is sometimes provided — axes must be labelled, including a scale if numerical data are given, the origin should be shown if appropriate and the general shape of the expected line should be drawn.
- **Explain** — use correct physics terminology and principles; the depth of your answer should reflect the number of marks available.
- **Show that** — usually a value is given so that you can proceed with the next part; you should show all your working and give your answer to more significant figures than the value given (to prove that you have actually done the calculation).
- **Calculate** — show all your working and give *units* at every stage; the number of significant figures in your answer should reflect the given data, but you should keep each stage in your calculator to prevent excessive rounding.
- **Determine** — means you will probably have to extract some data, often from a graph, in order to perform a calculation.
- **Estimate** — a calculation in which you have to make a sensible assumption, possibly about the value of one of the quantities — think, does this give a reasonable answer?
- **Suggest** — there is often no single correct answer; credit is given for sensible reasoning based on correct physics.
- **Discuss** — you need to sustain an argument, giving evidence for and against, based on your knowledge of physics and possibly using appropriate data to justify your answer.

You should pay particular attention to diagrams, sketching graphs and calculations. Many students lose marks by failing to label diagrams properly, by not giving essential numerical data on sketch graphs, by not showing all their working or by omitting the units in calculations.

About this section

The following two tests are made up of questions similar in style and content to the A-level physics examinations. The first test is in the style of A-level paper 1. The questions mainly concern the topics covered in this guide, but there are some complete or part questions requiring prior knowledge of core physics I topics.

You may like to attempt a complete paper in the allotted time and then check your answers, or maybe do the multiple-choice section and selected questions to fit your revision plan. It is worth noting that there are 90 marks available for the 105 minute test, so this should help in determining how long you should spend on a particular question. You should therefore be looking at about 10 minutes for the multiple-choice section and just over a minute a mark on the others.

Questions & Answers

The second test is in the style of A-level physics paper 3. Again, most of the questions are drawn from the topics covered in this guide (advanced physics I) but there will be some synoptic questions that may require prior knowledge from all the core physics material. This test will include questions relating to practical methods and techniques, including details from some of the core practicals.

You should also be aware that during the examination you must write your answers directly onto the paper. This will not be possible for the tests in this book, but the style and content are the same as the examination scripts in every other respect. It may be that diagrams and graphs that would normally be added to the paper have to be copied and redrawn. If you are doing a timed practice test, you should add an extra few minutes to allow for this.

The answers should not be treated as model answers because they represent the bare minimum necessary to gain marks. In some instances, the difference between an A-grade response and a C-grade response is suggested. This is not possible for the multiple-choice section, and many of the shorter questions do not require extended writing.

Ticks (✓) are included in the answers to indicate where the examiner has awarded a mark. Half marks are not given.

■Test paper 1

Time allowed: 1 hour 45 minutes. Answer **all** the questions.

Section A

For questions 1–10, select one answer from A to D.

Question 1

The angular velocity of a particle moving with a tangential velocity of $2.0\,\text{m}\,\text{s}^{-1}$ in a circular path of radius 20 cm is:

A $0.40\,\text{rad}\,\text{s}^{-1}$

B $0.80\,\text{rad}\,\text{s}^{-1}$

C $10\,\text{rad}\,\text{s}^{-1}$

D $20\,\text{rad}\,\text{s}^{-1}$ (1 mark)

Question 2

Which of the following is the same unit as the tesla?

A $N^{-1}Am$

B NAm

C NAm^{-1}

D $NA^{-1}m^{-1}$ (1 mark)

Question 3

Which of the following will *not* experience a resultant force?

A   an electron moving parallel to an electric field

B   an electron moving parallel to a magnetic field

C   an electron moving perpendicular to an electric field

D   a stationary electron in an electric field (1 mark)

Question 4

A muon has a mass of about $106\,\text{MeV}/c^2$. Its mass in kilograms is approximately:

A 2×10^{-34}

B 2×10^{-28}

C 2×10^{-25}

D 2×10^{-22} (1 mark)

ⓔ Remember to convert MeV to joules.

The following are four possible graphs of a quantity Y plotted against another quantity X. Refer to these graphs when answering questions 5, 6 and 7.

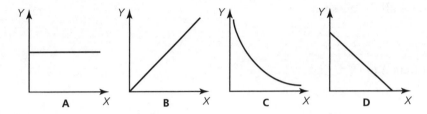

Question 5

Which graph *best* represents Y when it is the electric field strength close to a point charge and X is the distance from the charge? (1 mark)

Question 6

Which graph *best* represents Y when it is the force acting on a positively charged particle between two parallel plates with a constant potential difference across them and X is the distance from the positive plate? (1 mark)

Question 7

Which graph *best* represents Y when it is the momentum of a proton moving at right angles to a uniform magnetic field and X is the radius of the proton's path? (1 mark)

Question 8

The mains voltage in the UK is $230\,\text{V}_{\text{rms}}$. The peak value of the current in a heater of resistance R will be:

A $\dfrac{\sqrt{2}}{230R}$ A

B $\dfrac{230}{2R}$ A

C $\dfrac{230}{R\sqrt{2}}$ A

D $\dfrac{230\sqrt{2}}{R}$ A (1 mark)

Question 9

The momentum of an object of mass $0.12\,\text{kg}$ moving with kinetic energy $24\,\text{J}$ is:

A $2.4\,\text{kg}\,\text{m}\,\text{s}^{-1}$

B $4.8\,\text{kg}\,\text{m}\,\text{s}^{-1}$

C $5.8\,\text{kg}\,\text{m}\,\text{s}^{-1}$

D $12\,\text{kg}\,\text{m}\,\text{s}^{-1}$ (1 mark)

Question 10

A K⁻ meson is composed of which combination of quarks?

A us

B ūs

C dss

D uss

(1 mark)

Total: 10 marks

Answers to Questions 1–10

(1) C

ⓔ $\omega = \dfrac{v}{r} = \dfrac{2.0 \text{ m s}^{-1}}{0.20 \text{ m}} = 10 \text{ rad s}^{-1}$

(2) D

ⓔ This comes from $B = \dfrac{F}{Il}$

(3) D

ⓔ It is important to remember that an e.m.f. cannot be induced in a conductor unless there is a *change* in flux linkage — this requires a changing field at right angles to the conductor.

(4) B

ⓔ $m_0 = \dfrac{E_0}{c^2} = \dfrac{(106 \times 10^6 \text{ eV})(1.6 \times 10^{-19} \text{ J eV}^{-1})}{(3.0 \times 10^8 \text{ m s}^{-1})^2} = 1.9 \times 10^{-28} \text{ kg}$

(5) C

ⓔ The inverse square relationship $E = \dfrac{Q}{4\pi\varepsilon_0 r^2}$ is best represented by the curve in C.

(6) A

ⓔ This set up gives a uniform electric field. The field strength is constant, so the force ($F = Eq$) on the charged particle will be the same at all points between the plates.

(7) B

ⓔ The radius of the proton's path is given by $r = p/BQ$. Because the charge and the field strength are both constant, the momentum is directly proportional to the radius, and so the graph is a straight line through the origin.

(8) D

ⓔ The rms value of the voltage $\dfrac{V_0}{\sqrt{2}}$, so $V_0 = V_{\text{rms}}\sqrt{2}$ and $I_0 = \dfrac{V_0}{R} = \dfrac{230\sqrt{2}}{R}$

(9) A

ⓔ $E_k = \dfrac{p^2}{2m} \Rightarrow p = \sqrt{2mE_k} = \sqrt{2 \times 0.12 \text{ kg} \times 24 \text{ J}} = 2.4 \text{ kg m s}^{-1}$

(10) B

ⓔ A meson is made from a quark and an antiquark, so the only possible answer is B. You can check that the charges $-\frac{2}{3}e$ for ū and $-\frac{1}{3}e$ for s add up to $-1e$, which is correct for the K⁻ meson.

Questions & Answers

The specification does not require you to remember the quark composition of a K⁻ meson. However, you are expected to know that particles made up of three quarks are baryons; so even though dss also has a charge of –1e it cannot be the right answer.

Section B

Question 11

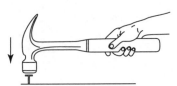

A hammer of mass 0.40 kg is used to drive a nail into a wooden board. The hammer head hits the nail at $6.2\,\text{m}\,\text{s}^{-1}$ and rebounds from the nail with a speed of $1.2\,\text{m}\,\text{s}^{-1}$. The hammer is in contact with the nail for 8.5 ms.

ⓔ This question is about change in momentum and the use of Newton's second law.

(a) Calculate the impulse of the hammer head on the nail. (2 marks)

ⓔ Momentum is a vector quantity. Remember to change the sign for motion in the opposite direction.

(b) Calculate the average force exerted by the hammer on the nail. (1 mark)

Total: 3 marks

Answer

(a) Taking downwards to be the positive direction:

impulse on hammer = change in momentum of hammer

$$= 0.40\,\text{kg} \times (-1.2\,\text{m}\,\text{s}^{-1}) - 0.40\,\text{kg} \times (+6.2\,\text{m}\,\text{s}^{-1})✓$$

$$= -3.0\,\text{kg}\,\text{m}\,\text{s}^{-1}$$

By Newton's third law, the hammer exerts an impulse of $+3.0\,\text{kg}\,\text{m}\,\text{s}^{-1}$ on the nail (i.e. downwards) ✓

(b) average force $= \dfrac{\text{impulse}}{\text{time}} = \dfrac{3.0\,\text{kg}\,\text{m}\,\text{s}^{-1}}{8.5\times10^{-3}\,\text{s}} = 350\,\text{N}$ ✓

Question 12

Beta-plus decay can be represented by the equation:

$$p \rightarrow n + e^+ + \nu_e$$

ⓔ This question tests the recognition of particle symbols and requires knowledge of the quark structures of protons and neutrons.

(a) Identify the four particles in the reaction, and state which are baryons and which are leptons. (3 marks)

(b) Use the laws of conservation of charge, baryon number and lepton number to check the validity of the reaction. (3 marks)

Total: 6 marks

> **Answer**
>
> (a) p is a proton; n is a neutron; these are baryons. ✓
>
> e⁺ is a positron and v_e is an electron neutrino; ✓; both are leptons ✓

e A grade-C student may lose a mark by just stating 'neutrino'. You are expected to be familiar with the fundamental particles and their symbols — including the three different kinds of neutrinos. Antiparticles should also be recognised (the term 'antielectron' would be accepted in place of the much more commonly used 'positron').

> (b) Charge, Q: $+1 \rightarrow 0 + (+1) + 0 = +1$ ✓

e It is important to write +1 or –1 (or +1e and –1e) rather than just '+' or '–' when representing the charge.

> Baryon number, B: $1 \rightarrow 1 + 0 + 0 = 1$ ✓
>
> Lepton number, L_e: $0 \rightarrow 0 + (-1) + 1 = 0$ ✓

e The positron is the antiparticle of the electron and so has a lepton number of –1.

Question 13

A gardener holds a wheelbarrow full of compost as shown.

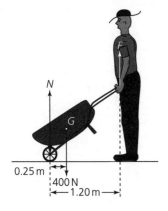

0.25 m
400 N
1.20 m

(a) Use the principle of moments to find the vertical force, F, exerted by the gardener on the handles of the wheelbarrow to keep it in equilibrium. (2 marks)

(b) Calculate the size of the normal force, N, of the ground on the wheel. (1 mark)

Total: 3 marks

Questions & Answers

> **Answer**
>
> **(a)** Taking moments about the axle:
>
> $400\,N \times 0.25\,m = F \times 1.20\,m$ ✓ $\rightarrow F = 83\,N$ ✓
>
> **(b)** For equilibrium:
>
> $F + N = 400\,N \rightarrow N = 317\,N$ ✓

Question 14

A light-dependent resistor (LDR) is used to control the switching of a mains lighting circuit. When the background illumination falls below a certain value, an electronic switch turns the mains lighting on using the potential divider circuit shown here.

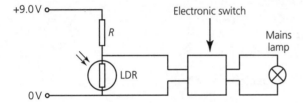

The variation of the resistance of the LDR with background illumination is shown in the graph.

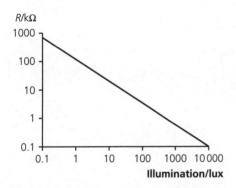

(a) What is the advantage of using a log–log graph for this LDR? (1 mark)

(b) With reference to charge carriers, explain why the resistance of the LDR changes with temperature. (2 marks)

The switch is adjusted so that it turns the mains lamps on when the potential difference across its input (V_0) reaches 6.0 V.

The circuit is designed so that the mains lamps will be switched on when the background illumination falls below 3.0 lux.

(c) Use the graph to determine the value of the resistance of the LDR when the illumination is 3.0 lux. (1 mark)

ⓔ Remember that the scale is not linear between the values given on the scale.

(d) Calculate the value of the resistor R if the circuit is used to switch at an illumination of 3.0 lux.

(2 marks)

Total: 6 marks

ⓔ You will need to use the potential divider equation derived in core physics I.

Answer

(a) A log–log scale is used to fit a wide range of values onto the graph. ✓

(b) The number of charge carriers reduces when the light intensity is lower ✓ leading to a higher resistance. ✓

(c) $R = 50\,k\Omega$ ✓

ⓔ A value of ±1 division is acceptable (40–60 kΩ). The unit must be correct.

(d) $V_0 = \dfrac{R_{LDR}}{R_{LDR} + R} \times 9\,V \Rightarrow \dfrac{6.0\,V}{9.0\,V} = \dfrac{50k\Omega}{50k\Omega + R}$ ✓

$R = 25\,k\Omega$ ✓

ⓔ These marks will be given if an incorrect value for the resistance of the LDR is used.

Question 15

A fairground ride includes a loop in the form of a vertical circle of radius 5.0 m.

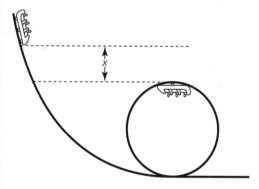

ⓔ This question is about circular motion and energy conservation.

(a) Show that the minimum speed of the cart needed to ensure that the wheels stay in contact with the track at the top of the loop is $7\,m\,s^{-1}$.

(2 marks)

ⓔ At the top of the loop the centripetal force will be vertically down and is generally provided by the weight of the cart and the reaction force of the track on the wheels.

Questions & Answers

(b) Calculate the minimum height, x, above the top of the loop through which the cart must fall to complete the loop.

(2 marks)

(c) In practice, the cart always falls from a height that is much bigger than that calculated in part (b). Explain why this is necessary.

(1 mark)

Total: 5 marks

Answer

(a) The force of the track on the wheels will be zero and so the centripetal force is provided by the weight:

$$\frac{mv^2}{r} = mg \checkmark$$

$$v = \sqrt{gr} = \sqrt{9.8\,\mathrm{m\,s^{-2}} \times 5.0\,\mathrm{m}} = 7.0\,\mathrm{m\,s^{-1}} \checkmark$$

ⓔ This is a 'show that' question so the answer must be given to one more significant figure.

(b) Loss in gravitational potential energy = gain in kinetic energy

$$mgx = \tfrac{1}{2}mv^2$$

$$x = \frac{v^2}{2g} = \frac{\left(7.0\mathrm{ms^{-1}}\right)^2}{\left(2 \times 9.8\mathrm{ms^{-2}}\right)} \checkmark$$

$$= 2.5\,\mathrm{m} \checkmark$$

(c) Some of the gravitational potential energy will be converted into other forms of energy because of friction, air resistance etc. ✓

Question 16

In an experiment to investigate the charge on an electron, Robert Millikan observed charged oil droplets in the electric field between a pair of charged parallel plates.

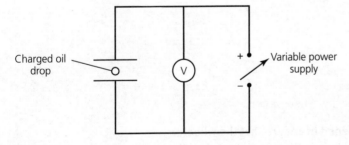

ⓔ This question is about uniform electric fields. You need to apply the basic equation that defines electric field strength ($E = F/Q$) and that for the field strength between a pair of parallel plates ($E = V/d$)

(a) Draw a diagram to show the pattern of the electric field between the plates. (2 marks)

A droplet is seen moving upwards between the plates. The voltage is adjusted so that the droplet stops moving and is held stationary in the field.

(b) Draw a free-body diagram for a charged droplet in the field. Label all the forces acting on the droplet. (2 marks)

Given that the mass of the droplet is 9.8×10^{-16} kg, the separation of the plates is 0.80 cm and the potential difference across the plates needed to hold the droplet stationary in the field is 240 V, calculate:

(c) the electric field strength between the plates (2 marks)

(d) the magnitude of the charge on the oil drop (3 marks)

Total: 9 marks

Answer

(a) The diagram should have:

- lines drawn between the parallel plates, at right angles to the plates ✓
- equal spacing between consecutive lines and arrows on the lines pointing from positive to negative ✓

ℹ Many students lose marks by drawing irregularly spaced lines or by omitting the direction. It is important to remember that a uniform field must be represented by evenly spaced parallel lines.

(b) The diagram should include:

- an upward force labelled 'electric force' or Eq ✓
- a downward force labelled 'weight', W or mg ✓

(c) $E = \dfrac{V}{d} = \dfrac{240\,\text{V}}{0.80 \times 10^{-2}\,\text{m}}$ ✓ $= 3.0 \times 10^{4}\,\text{V m}^{-1}$ ✓

(d) $Eq = mg$ ✓ so $q = \dfrac{mg}{E} = \dfrac{(9.8 \times 10^{-16}\,\text{kg}) \times 9.8\,\text{ms}^{-2}}{3 \times 10^{4}\,\text{N C}^{-1}}$ ✓ $= 3.2 \times 10^{-19}\,\text{C}$ ✓

Question 17

A car approaches a junction in icy conditions. The driver brakes and skids to a halt, but is unable to avoid an oblique collision with another car.

The moving car, of mass 1500 kg, tries to brake but also skids, and is travelling at 10 m s^{-1} at the instant the collision occurs. It deflects off the side of the stationary vehicle at an angle of 20° to its initial direction. The stationary car, of mass 2000 kg, is shunted with an initial velocity of 4.0 m s^{-1} at 60° in the other direction, as shown in the diagram.

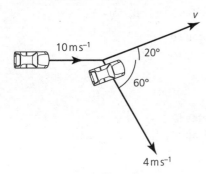

e The law of conservation of momentum in two dimensions is needed here, together with the application of Newton's second law of motion.

(a) If the velocity of the deflected car is v, write down the components, in the original direction of the moving car, of the linear momentum of each car immediately after impact. (2 marks)

(b) Show that v is about $8\,\text{m s}^{-1}$. (2 marks)

(c) If the duration of the impact is $0.25\,\text{s}$, calculate the average force on the side of the stationary car. (2 marks)

(d) Modern cars have 'side impact protection systems' (SIPS) that allow the doors to crumple on impact. Explain how this system can reduce the effects of a side collision on the occupants of a car involved in an accident. (2 marks)

Total: 8 marks

Answer

(a) The components are $1500\,\text{kg} \times v\cos 20°$ ✓ and $2000\,\text{kg} \times 4.0\,\text{m s}^{-1} \times \cos 60°$. ✓

(b) Using the principle of conservation of linear momentum:

$(1500\,\text{kg} \times 10\,\text{m s}^{-1}) + 0\,\text{kg m s}^{-1} = (1500\,\text{kg} \times v\cos 20°) + (2000\,\text{kg} \times 4.0\,\text{m s}^{-1} \times \cos 60°)$ ✓ which gives $v = 7.8\,\text{m s}^{-1}$ ✓

(c) average force $= \dfrac{\text{change in momentum}}{\text{time}}$ in the direction that the car is shunted ✓

$= \dfrac{(2000\,\text{kg} \times 4.0\,\text{m s}^{-1}) - 0\,\text{kg m s}^{-1}}{0.25\,\text{s}} = 32\,\text{k}$ ✓

e A grade-A student will realise that the force of impact on the stationary car must be in the direction the car is shunted. A grade-C student may use the component in the initial direction of the moving car, and so gain just one of the marks.

(d) The crumple zone extends the duration of the impact ✓ and so reduces the resultant force on the car. ✓

Question 18

A capacitor is charged using a 12 V battery and then discharged through a 220 kΩ resistor. The discharge current is measured at regular intervals and a graph of current against time is plotted.

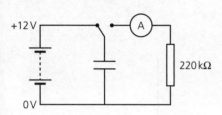
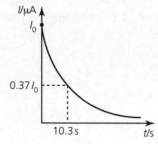

e This question relates to the exponential discharge of a capacitor through a resistor (core practical 11).

(a) Determine the initial value of the current, I_0. (1 mark)

(b) How could you use the graph to estimate the initial charge stored on the capacitor? (1 mark)

(c) The time taken for the current to fall to 37% of the initial value is equal to the time constant, and is found to be 10.3 s. Define the term *time constant* for a capacitor discharging through a resistor, and calculate the value of the capacitance of the capacitor. (2 marks)

(d) Calculate the value of I after 20.6 s. (2 marks)

The resistor in the circuit is now replaced by an electric motor, and a 10 000 μF capacitor is used in place of the original one. When the capacitor is fully charged and then discharged through the motor, a load of 20 g is raised through a height of 80 cm by the motor.

(e) Calculate the energy stored in the 10 000 μF capacitor when it is fully charged. (2 marks)

e The data sheet gives the equation $E = \frac{1}{2}QV$ for the energy stored in a capacitor. You will need to use the definition of capacitance to express this in terms of C and V.

(f) Calculate the work done by the motor in raising the load, and determine the efficiency of the system. (3 marks)

Total: 11 marks

Answer

(a) $I_0 = \dfrac{V_0}{R} = \dfrac{12\ V}{220 \times 10^3\ \Omega} = 5.5 \times 10^{-5}\ A = 55\ \mu A$ ✓

(b) Q = area under the I–t graph ✓

(c) The time constant for the circuit is RC. ✓ Here we have:

$220 \times 10^3\ \Omega \times C = 10.3\ s$, so $C = \dfrac{10.3\ s}{220 \times 10^3\ \Omega} = 4.7 \times 10^{-5}\ F = 47\ \mu F$ ✓

(d) 20.6 s is equal to two time constants. ✓

In 10.3 s, I falls to 0.37 × 55 μA and in the next 10.3 s it falls to 0.37 × (0.37 × 55 μA) = 7.5 μA. ✓

@ An equally valid method is to substitute $t = 20.6$ s into the formula $I = I_0 e^{-t/RC}$. This would give the answer 7.4 μA.

(e) $E = \frac{1}{2}CV^2 = \frac{1}{2} \times (10\,000 \times 10^{-6}\,\text{F}) \times (12\,\text{V})^2$ ✓ = 0.72 J ✓

(f) $W = mgh = 0.020\,\text{kg} \times 9.8\,\text{m s}^{-2} \times 0.80\,\text{m} = 0.16\,\text{J}$ ✓

efficiency $= \dfrac{0.16\,\text{J}}{0.72\,\text{J}} \times 100\%$ ✓ = 22% ✓

Question 19

Linear accelerators are used to produce high-energy particles. The diagram shows the structure of part of such a device.

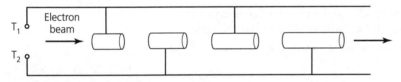

@ This question requires a basic knowledge of a linear accelerator.

(a) State the nature of the power supply connected across terminals T_1 and T_2. (2 marks)

(b) Explain why the electrons travel with constant velocity when they are inside the tubes. (2 marks)

(c) Explain why the tubes increase in length along the accelerator. (2 marks)

(d) The Stanford linear accelerator is capable of accelerating electrons to energies around 30 GeV. Convert this energy into joules. (1 mark)

(e) State *two* reasons why such high-energy particles are needed for investigating fundamental particles. (2 marks)

Total: 9 marks

Answer

(a) An alternating supply of high voltage ✓ and high frequency. ✓

@ To gain both marks it must be clear that the potential difference (voltage) is alternating. Writing 'an AC supply' only is insufficient.

(b) There is no electric field inside the tubes (every part of each tube is at the same voltage at any instant), ✓ so no force acts on the electrons while they are inside a tube. ✓

(c) The speed of the electrons increases along the accelerator. ✓ Therefore the tubes need to get longer so that the time the electrons spend in each stays the same. ✓

(d) $30\,\text{GeV} = 30 \times 10^9\,\text{eV} \times 1.6 \times 10^{-19}\,\text{J eV}^{-1} = 4.8 \times 10^{-9}\,\text{J}$ ✓

(e) Any *two* of the following reasons:

- High energies are needed to overcome repulsive forces or to break up particles into their constituents. ✓
- Sufficient energy is required for the rest-mass energy of new particles created. ✓
- Higher energies mean shorter wavelengths (from $E = hc/\lambda$) and short wavelengths are needed to examine fine structure. ✓

Question 20

A petrol-engine ignition system uses an induction coil to generate the high voltage needed to create sparks across the electrodes of a spark plug to ignite the fuel–air mixture in the cylinders. A schematic diagram is shown.

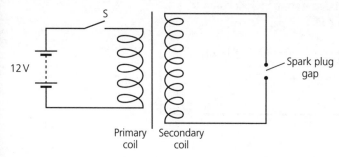

ⓔ This question is about electromagnetic induction. When the current is switched off in the primary coil, the changing flux is linked with the secondary coil.

(a) For a spark to jump between the electrodes, the electric field strength needs to be $3.0 \times 10^7\,\text{V m}^{-1}$. Show that, for a gap of 0.7 mm, the voltage across the plug is about 20 kV.

(2 marks)

To produce such high voltages, the primary circuit is broken so that the current falls rapidly to zero.

(b) Use Faraday's law to explain how this leads to a high voltage being induced in the secondary coil.

(4 marks)

ⓔ The question specifically asks for the use of Faraday's law, so you cannot gain full marks unless you include a statement of the law in your answer.

(c) The secondary coil in one system has 20 000 turns and a cross-sectional area of $4.0 \times 10^{-3}\,\text{m}^2$. The flux density linked with the coils is 1.2 T, which drops to zero in 3.6 ms when the circuit of the primary coil is broken. Calculate:

(i) the flux linkage of the secondary coil before the circuit is broken (1 mark)

(ii) the voltage induced across the ends of the secondary coil (2 marks)

Total: 9 marks

Answer

(a) $E = \dfrac{V}{d} \Rightarrow V = Ed$ ✓ $= (3.0 \times 10^7\,\text{V}\,\text{m}^{-1}) \times (0.7 \times 10^{-3}\,\text{m}) = 2.1 \times 10^4\,\text{V} = 21\,\text{kV}$ ✓

(b) Faraday's law states that the induced e.m.f. is proportional to the rate of change of flux linkage in a conductor. ✓ When the primary circuit is broken, the magnetic flux due to the current in the primary coil falls rapidly to zero. ✓ This causes a rapid change in the flux linkage with the secondary coil, ✓ which has a large number of turns, ✓ and so a large e.m.f. is induced in the secondary coil.

@ One example of a poor response is:

'When S is closed a field is formed in the primary coil. When the switch is opened the field drops to zero and lines cut through the secondary coil and produce a large current in it.' (0/4)

The student has not mentioned *magnetic* flux. Although cutting field lines describes the process, *change in flux linkage* must be stated, and according to Faraday's law, an *e.m.f. is induced* in the secondary coil.

(c) (i) flux linkage $= N\phi = NBA = (2.0 \times 10^4) \times 1.2\,\text{T} \times (4.0 \times 10^{-3}\,\text{m}^2) = 96\,\text{Wb}$ ✓

(ii) $\varepsilon = -\dfrac{\text{d}(N\phi)}{\text{d}t}$ ✓ $= \dfrac{(96 - 0)\,\text{Wb}}{3.6 \times 10^{-3}\,\text{s}} = 27\,\text{kV}$ ✓

Question 21

(a) Outline the alpha-particle scattering experiment used by Geiger and Marsden to establish the nuclear model of the atom. You should include details of the results of the experiment and the conclusions drawn from them. (5 marks)

The symbols for an alpha particle and a gold nucleus are ^4_2He and $^{197}_{79}\text{Au}$, respectively.

(b) Explain the meaning of the numbers on the symbols. (2 marks)

(c) Show that the mass of an alpha particle is about $7 \times 10^{-27}\,\text{kg}$. (1 mark)

@ This is a 'show that' question — so it needs an answer to one significant figure more than quoted.

(d) The radius of a gold nucleus is $6.8 \times 10^{-15}\,\text{m}$. The alpha particles used for the scattering experiment have sufficient energy to directly approach to within ten times this distance from the nucleus. Calculate the force between a gold nucleus and an alpha particle at this separation. (3 marks)

Total: 11 marks

Total for paper = 90 marks

e Coulomb's law should be applied.

Answer

(a) Alpha particles are fired at a very thin gold foil in an evacuated chamber. Although most of the alpha particles pass straight through the foil, some are deviated through a small angle. An extremely small number are deflected through large angles or straight back. These results suggest that the atom is mostly empty space and that most of its mass is concentrated at its centre in a positively charged nucleus.

e This answer would be awarded the full 5 marks. Any five of the following points would be accepted:

- Alpha particles are fired at a (thin) gold film in a vacuum. ✓
- Most particles are not deflected and pass straight through the foil. ✓
- Some particles are deflected. ✓
- Very few particles are deviated through large angles (or are reflected straight back). ✓
- These observations suggest that atom is mostly empty space ✓ and that the mass of each gold atom is concentrated in a very small nucleus ✓ which carries a positive charge. ✓

(b) The lower number represents the number of protons in the nucleus (the proton number) and the upper one is the total number of protons and neutrons (the nucleon number). ✓

So the alpha particle contains two protons and two neutrons; the gold nucleus contains 79 protons and 118 neutrons. ✓ (Either of these two examples would earn the mark.)

e Since the question says 'explain', to gain both marks reference must be made to at least one of the given symbols.

(c) mass $\approx 4 \times 1.67 \times 10^{-27}$ kg (neutrons have a similar mass to protons)

$= 6.7 \times 10^{-27}$ kg ✓

(d) $F = \dfrac{Q_1 Q_2}{4\pi\varepsilon_0 r^2}$ ✓

$= \dfrac{(2 \times 1.6 \times 10^{-19}\,\text{C}) \times (79 \times 1.6 \times 10^{-19}\,\text{C})}{4\pi \times (8.85 \times 10^{-12}\,\text{F m}^{-1}) \times (10 \times 6.8 \times 10^{-15}\,\text{m})^2}$ ✓

$= 7.9\,\text{N}$ ✓

■ Test paper 2

Time allowed: 2 hours 30 minutes. Answer **all** the questions.

Question 1

A student is asked to measure the diameters of a test tube and of a piece of string as accurately as he can. He is given only about a 1 metre length of string that has an approximate diameter of 1 mm, a test tube with a diameter of about 2 cm and a metre rule.

ⓔ You are expected to select a method for both measurements that would give the least uncertainty, using just the metre rule.

Describe the measurements he should take and estimate the percentage uncertainty in the value of each diameter.

(6 marks)

Total: 6 marks

ⓔ To gain the marks you need to use techniques for measuring lengths that are considerably larger than a direct measurement.

> ### Answer
>
> *Diameter of tube:* wrap the string tightly around the tube as many times as possible. ✓
>
> Measure the length of a complete number of turns. ✓ The diameter is found by dividing the length by the number of turns. ✓
>
> 1.00 m of string will wrap around a tube of diameter of about 2 cm about 15 times. If 10 turns are used, $l \approx 63$ cm.
>
> The percentage uncertainty will be $\pm\dfrac{1\,\text{mm}}{630\,\text{mm}} \times 100\% \approx \pm 0.2\% \approx \pm\, 0.2\%$ ✓

ⓔ The resolution of the scale on a metre rule is usually taken as ±0.5 mm but marking the ends of the string would make such precision difficult in this case. The mark would be awarded for either value.

> *Diameter of string:* for 10 turns the length of the coil will be about 10 mm. ✓
>
> The percentage uncertainty $= \pm\dfrac{0.5\,\text{mm}}{10\,\text{mm}} \times 100\% \approx \pm 5\%$ ✓

Question 2

A bar magnet is dropped through a coil of copper wire, which is connected to the input of a data-logging device. The data logger displays a voltage–time graph as shown below:

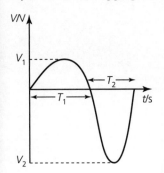

Explain the shape of the graph, making reference to the relative values of V_1 and V_2 and of T_1 and T_2.

(3 marks)

Total: 3 marks

Answer

When the magnet leaves the coil, it is moving faster than it was when entering the coil, so the induced e.m.f. is higher, i.e. the magnitude of V_2 is bigger than the magnitude of V_1. ✓

As the magnet leaves the coil, the direction of the induced e.m.f. is reversed, so V_1 and V_2 have opposite signs. ✓

The magnet accelerates as it drops through the coil, so it takes less time to exit than to enter; therefore the duration T_2 will be shorter than the duration T_1. ✓

ⓔ A grade-C student may describe the induction process adequately but fail to specifically compare the values or polarities of the voltages and times.

Question 3

A student drops a ball from a height of 2.50 m above the floor. Two other students use stopclocks to measure the time taken for the ball to strike the floor. They repeat the timings four more times — they record their data as shown in the table.

	Time/s	Mean time/s
Student A	0.71, 0.74, 0.70, 0.88, 0.72	0.75
Student B	0.70, 0.73, 0.72, 0.74, 0.72	0.72

ⓔ This question relates to core practical 1. You should be aware of experimental techniques and practices used for this exercise.

(a) Explain why the mean value for the time stated by student A is likely to be less reliable than that of student B.

(2 marks)

(b) Determine the percentage uncertainty in student B's times.

(2 marks)

(c) Use the mean value of the times taken by student B to calculate a value for the acceleration due to gravity. [2 marks]

A more reliable method for measuring the acceleration due to gravity uses light gates to measure the time for a ball bearing to fall through a range of distances.

(d) Describe how the results of such an experiment could be used to plot a graph from which the value of g can be obtained. Details of how the acceleration due to gravity is determined from the graph should be given. [3 marks]

Total: 9 marks

Answer

(a) Student A's values include an anomalous reading (0.88 s), i.e. one that is very much different to the others. ✓

This should not have been included when calculating the mean value. ✓

(b) mean value of time = 0.72 ± 0.02 s ✓

percentage uncertainty $= \dfrac{\pm 0.02\,\text{s}}{0.72\,\text{s}} \times 100\% = \pm 3\%$ ✓

(c) Using $s = ut + \frac{1}{2}gt^2$, when $u = 0$ $g = \dfrac{2h}{t^2} = \dfrac{2 \times 2.50\,\text{m}}{(0.72\,\text{s})^2}$ ✓ $= 9.6\,\text{m s}^{-2}$ ✓

(d) Plot a graph of h/m against t^2/s^2 ✓; measure the gradient ✓

Gradient = $\frac{1}{2}g$ $g = 2 \times$ gradient ✓

Question 4

A capacitor is charged to a potential difference of 6.0 V as shown in this circuit:

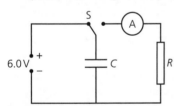

The switch is thrown so that the capacitor discharges through the ammeter and the resistor R.

ⓔ This question is based on the experimental procedures used in core practical 11.

The ammeter reading is taken every 10 s for the first minute during the discharge — the results are shown in the table.

t/s	$I/\mu\text{A}$	$\ln(I/\mu\text{A})$
0	12.8	
10	10.3	
20	8.4	
30	6.8	
40	5.5	
50	4.4	
60	3.6	

(a) Calculate the value of R (assume that the resistance of the ammeter is negligible). Give your answer in kΩ.

(1 mark)

The current in the circuit at a time t is given by the equation $I = I_0 e^{-t/RC}$.

e You should be aware that the equation can be written as $\ln I = \ln I_0 - t/CR$

(b) Complete the table to show the values of $\ln(I/\mu A)$.

(1 mark)

(c) Plot a graph of $\ln(I/\mu A)$ against t/s on the grid provided.

(2 marks)

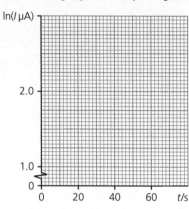

(d) Use the graph to determine the time constant RC of the circuit, and so find the value of the capacitance of the capacitor.

(3 marks)

Total: 7 marks

Answer

(a) $R = \dfrac{V_0}{I_0} = \dfrac{6.0\,\text{V}}{12.8 \times 10^{-6}\,\text{A}} = 4.7 \times 10^5\,\Omega = 470\,\text{k}\Omega$ ✓

(b) $\ln(I/\mu A)$ 2.55   2.33   2.13   1.92   1.70   1.48   1.28 ✓

(c)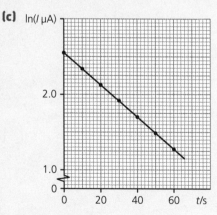

Plots ✓; line ✓

(d) Gradient = $0.021\,\text{s}^{-1}$ ✓

$= \dfrac{1}{RC} \rightarrow RC = 47\,\text{s} \rightarrow C = \dfrac{47\,\text{s}}{4.7 \times 10^5\,\Omega} = 1.0 \times 10^{-4}\,\text{F} = 100\,\mu\text{F}$ ✓✓

Question 5

An experiment to investigate the law of conservation of momentum is set up as shown in the diagram.

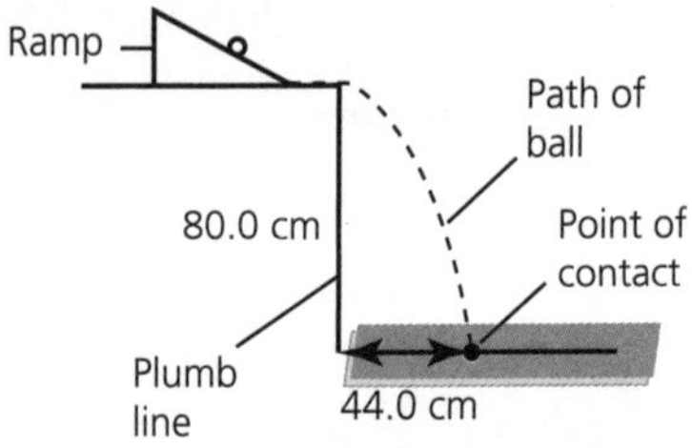

A steel ball bearing of mass 46 g, is released from a marked position on the ramp and projected across the edge of the bench. When it strikes the carbon paper it leaves a mark on the plain paper beneath it. The horizontal distance from the tip of the plumb line (vertically below the point of projection) to the mark is found to be 44.0 cm.

The vertical height of the bench surface above the floor is 80.0 cm

(a) Show that:

(i) the time taken for the ball to fall to the floor is approximately 0.4 s

(ii) the horizontal velocity of the ball as it leaves the bench is approximately 1 m s^{-1} (3 marks)

e You will need to apply the equations of motion to the vertical and horizontal components of the motion.

An identical ball is placed on the edge of the bench and the first ball is released from the same mark on the ramp as before. The ramp has been adjusted so that the moving ball strikes the stationary one obliquely, and both balls fall onto the carbon paper as shown in the diagram.

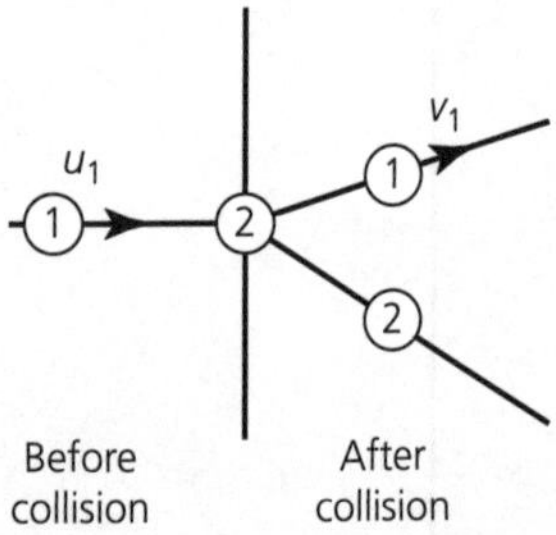

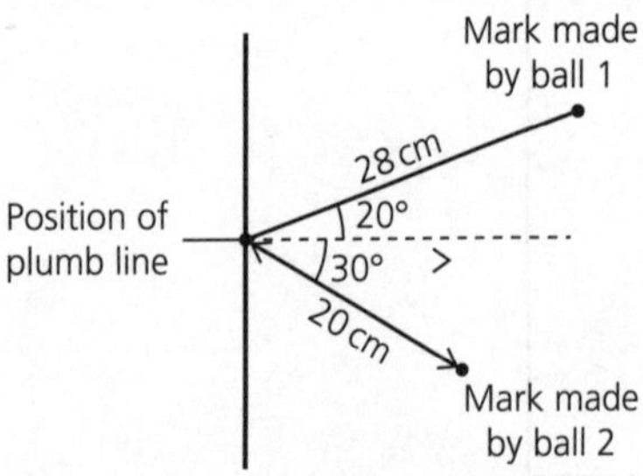

(b) Use the measurements of the positions of the marks on the paper to draw a vector diagram to represent the vector sum of the momentum of the two balls after the collision. Comment on the extent to which the collision obeys the law of conservation of linear momentum. (5 marks)

ⓔ Because the masses of the balls are identical, a vector diagram for the velocities is acceptable.

(c) State, with a reason, whether the collision between the balls is elastic or inelastic. (2 marks)

A student has downloaded an app for her smartphone that enables the camera to take still frames at 100 frames per second and display the outcome as a single picture.

(d) Outline a method by which she could use the phone to investigate the law of conservation of linear momentum for an oblique collision between a ball bearing moving along a surface and a stationary one. You should include a sketch of the images before and after the impact, and an explanation of how measurements from the sketch can be used to check the validity of the law. (6 marks)

Total: 16 marks

ⓔ This is one of the methods suggested for the study of oblique collisions for **core practical 10**.

Answer

(a) (i) Use $s = ut + \frac{1}{2}at^2 \rightarrow t = \sqrt{\frac{2s}{g}} = \sqrt{\frac{2 \times 0.800\,\text{m}}{9.8\,\text{m s}^{-2}}} = 0.40\,\text{s}$ ✓✓

(ii) horizontal velocity = $\dfrac{\text{horizontal distance}}{\text{time}} = \dfrac{0.44\,\text{m}}{0.40\,\text{s}} = 1.1\,\text{m s}^{-1}$ ✓

(b) horizontal velocity of ball 1 $= \dfrac{0.28\,\text{m}}{0.40\,\text{s}} = 0.70\,\text{m s}^{-1}$

horizontal velocity of ball 2 $= \dfrac{0.20\,\text{m}}{0.40\,\text{s}} = 0.50\,\text{m s}^{-1}$ ✓ (either value)

or momentum of ball 1 = $0.032\,\text{kg m s}^{-1}$

momentum of ball 2 = $0.023\,\text{kg m s}^{-1}$

Scale: $1.0\,\text{cm} = 0.010\,\text{kg m s}^{-1}$

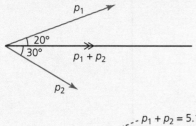

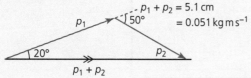

$p_1 + p_2 = 5.1\,\text{cm}$
$= 0.051\,\text{kg m s}^{-1}$

Plus any four from:

- vector diagram drawn (either triangle or parallelogram) ✓
- scales shown corresponding to either 20 cm and 28 cm, 0.50 m s^{-1} and 0.70 m s^{-1} or 0.023 kg m s^{-1} and 0.032 kg m s^{-1} ✓
- angle = 50° ✓
- resultant = 44 cm or 1.1 m s^{-1} or 0.051 kg m s^{-1} ✓
- comment: the vector sum of the momentum before the collision = the vector sum of the momentum after the collision with values to justify ✓

ⓔ All vector diagrams should be drawn accurately using a ruler and pencil and with the scale shown. Arrows should be drawn and labelled clearly.

(c) KE before collision = $\frac{1}{2}$ × 0.046 kg × (1.1 m s^{-1})^2 + 0
$$= 0.028\,J$$
KE after collision = $\frac{1}{2}$ × 0.046 kg × (0.7 m s^{-1})^2
$$= 0.017\,J \checkmark$$
As KE has been lost, the collision is inelastic ✓

ⓔ To get the marks, the answer must be supported by relevant data.

(d) Any six of the following:

- Line up a ramp on a level surface so that a ball rolls down the ramp and collides obliquely with an identical, stationary ball. ✓
- Illuminate with a powerful lamp and clamp the phone/camera above the region of the collision. ✓
- Release the first ball and activate the camera in multiframe mode. ✓
- Repeat several times. ✓
- Diagram showing images of balls before and after the collision. ✓

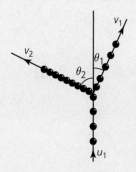

- State how u, v_1 and v_2 are calculated. ✓
- Measure θ_1 and θ_2. ✓
- Show that: $u = v_1 \cos \theta_1 + v_2 \cos \theta_2$. ✓

Question 6

The table includes four fundamental particles. For each particle, indicate its nature by ticking the appropriate column or columns.

(4 marks)

Particle	Baryon	Hadron	Lepton
Neutron			
Positron			
π^+ meson			
Tau neutrino			

Total: 4 marks

ℯ This question requires a basic knowledge of fundamental particles.

> **Answer**
>
> neutron: *both* baryon *and* hadron ✓
>
> positron: lepton *only* ✓
>
> π^+ meson: hadron *only* ✓
>
> tau neutrino: lepton *only* ✓

Question 7

To investigate the force between charged objects, two identical polystyrene spheres attached to highly insulating Perspex rods are set up as shown in the diagram. The lower sphere is placed on a sensitive top-pan balance and the other one is held vertically above it. The separation, x, of the spheres is measured from the shadows cast by a lamp onto a screen with a scale marked on it.

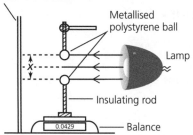

The balance is zeroed and then both spheres are charged to the same high voltage using a van de Graaff generator. The force that each sphere exerts on the other is recorded from the reading in grams on the balance.

The process is repeated for a range of values of r and the table shows a set of measurements that is obtained

r/m	0.100	0.150	0.200	0.250	0.300	0.350
mass/g	0. 092	0.041	0.0240	0.015	0.009	0.007
F/N						
log(r/m)	−1.00	−0.82	−0.70	−0.60	−0.52	−0.46
log(F/N)						

(a) Copy and complete the table by adding the values for F/N and log(F/N). (2 marks)

Coulomb's law can be written in the form $F = kr^n$, where k and n are constants.

(b) Give an expression for k and a value for n. What assumption needs to be made for the experiment to be valid? (3 marks)

A graph of log(F/N) against log(r/m) has been drawn using the measurements given in the table.

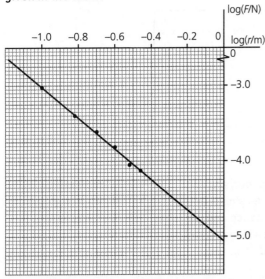

(c) Explain how k and n can be found from a log F against r graph and use the graph to determine their values. (5 marks)

(d) Calculate the size of the charge on each sphere. (2 marks)

(e) Why is it recommended that a hairdryer is used to keep the air around the spheres dry throughout the experiment? (1 mark)

Total: 13 marks

Answer

(a)

F/N	9.03×10^{-4}	4.02×10^{-4}	2.35×10^{-4}	1.47×10^{-4}	8.83×10^{-5}	6.87×10^{-5}
log(F/N)	−3.04	−3.40	−3.63	−3.83	−4.05	−4.16

✓✓

(b) $F = \dfrac{Q_1 Q_2}{4\pi\varepsilon_0 r^2} = \dfrac{Q_1 Q_2}{4\pi\varepsilon_0} \times r^{-2}$

$k = \dfrac{Q_1 Q_2}{4\pi\varepsilon_0}$ ✓ and n = −2 ✓

It is assumed that Q_1 and Q_2 remain the same throughout the experiment (no charge is lost from the spheres). ✓

(c) $\log F = \log(kr^n) = \log k + n \log r$ ✓

So a graph of $\log F$ against $\log r$ will have a gradient of n ✓ and an intercept of $\log k$ on the y-axis. ✓

From the graph: gradient $= -\dfrac{2.05}{1.02} = -2.0$ ✓ and the intercept $= -5.05$ ✓

(d) Because $Q_1 = Q_2$, $\log \dfrac{Q^2}{4\pi\varepsilon_0} = -5.05$ and $\dfrac{Q^2}{4\pi\varepsilon_0} = 8.91 \times 10^{-6}\,\text{Nm}^2$ ✓

$Q = \sqrt{8.85 \times 10^{-12}\,\text{N}^{-1}\text{m}^{-2}\text{C}^2 \times 8.91 \times 10^{-6}\,\text{Nm}^2} = 3.1 \times 10^{-8}\,\text{C}$ ✓

(e) Dry air reduces the chances of charge leaking to the surroundings. ✓

Question 8

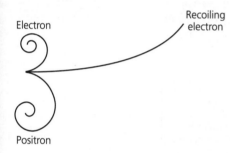

Electron

Recoiling electron

Positron

The diagram shows the tracks in a bubble chamber produced when a high-energy gamma ray interacts with a stationary electron to create an electron–positron pair.

This reaction is represented by the equation:

$$\gamma \rightarrow e^- + e^+$$

The third track is that of the originally stationary electron recoiling.

ⓔ This is a typical image of a bubble chamber event. The curvature of the tracks in the magnetic field is related to the momentum, and hence the velocity, of the particle, and the charge can be deduced using Fleming's left-hand rule.

(a) Why is there no trace for the gamma-ray photon? (1 mark)

(b) Explain how the laws of conservation of charge and energy apply to this interaction. (4 marks)

(c) What other conservation law must apply to this event? (1 mark)

(d) The magnetic field across the chamber is perpendicular to the plane of the diagram. Is the direction of the field into or out of plane of the paper? (1 mark)

(e) Explain how you could deduce that the speed of the recoiling electron is higher than the speeds of the electron and positron produced by the reaction. (1 mark)

(f) Calculate the maximum wavelength of the gamma ray that can create an electron–positron pair, given that the rest mass of an electron is $0.512\,\text{MeV}/c^2$. (4 marks)

Total: 12 marks

ⓔ You need to convert the rest mass to joules and use $E = hc/\lambda$. The positron is the antiparticle of the electron and so will have the same mass.

> **Answer**
>
> **(a)** The gamma ray is non-ionising (the photons carry no charge), and so does not leave a bubble trail. ✓
>
> **(b)** Conservation of charge:
>
> initial charge carried by γ and $e^- = 0 + (-1e) = -1e$
>
> final charge carried by e^-, e^+ and recoiling $e^- = (-1e) + (+1e) + (-1e) = -1e$ ✓

ⓔ An answer that compares the initial charge carried by the incident gamma-ray photon (0) and the final charge carried by the electron–positron pair $((-1e) + (+1e) = 0)$ would also be accepted. It is also acceptable to omit the symbol e, leaving $0 + (-1) = (-1) + (+1) + (-1)$.

> Conservation of energy:
>
> initial energy of photon = hf ✓
>
> final energy = rest-mass energy of positron + rest-mass energy of electron ✓ + kinetic energy of electron–positron pair and recoil electron ✓
>
> **(c)** Momentum must also be conserved. ✓
>
> **(d)** By Fleming's left-hand rule, the magnetic field is out of the paper. ✓

ⓔ Conventional current flow is indicated by the direction of the positron.

> **(e)** All three electrons have the same mass and the path of the recoiling electron has a larger radius than the trails of the electron and positron. ✓
>
> $$r = \frac{p}{Bq} = \frac{mv}{Bq}$$
>
> **(f)** Minimum energy needed to create an electron–positron pair is:
>
> $2 \times 0.512\,\text{MeV} = 1.024\,\text{MeV}$ ✓
>
> $= (1.024 \times 10^6\,\text{eV}) \times (1.6 \times 10^{-19}\,\text{J eV}^{-1})$ ✓
>
> $= 1.64 \times 10^{-13}\,\text{J}$
>
> energy of photon $= \dfrac{hc}{\lambda}$ ✓
>
> $\lambda_{max} = \dfrac{hc}{E_{min}} = \dfrac{(6.63 \times 10^{-34}\,\text{Js}) \times (3.00 \times 10^8\,\text{ms}^{-1})}{1.64 \times 10^{-13}\,\text{J}} = 1.2 \times 10^{-12}\,\text{m}$ ✓

Question 9

The displacement–time graph represents the motion of a large, soft ball of mass 400 g falling from a height of 1 metre and bouncing once.

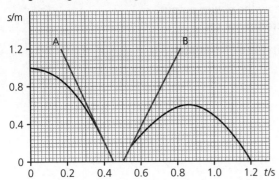

The lines A and B are the gradients of the graph at the points when the ball first strikes the floor, and when it leaves to start the upward part of the bounce.

(a) Use the lines to determine the velocity of the ball when it hits the floor and when it leaves the floor. (3 marks)

Ⓔ Remember that velocities are vector quantities and require a positive or negative sign to indicate the direction. It is quite difficult to determine the gradients from such a small graph. Make sure that you use the biggest triangle you can to find the gradient as accurately as possible.

(b) Calculate the magnitude of:

(i) The impulse of the floor on the ball.

(ii) The average force that the floor exerts on the ball during the bounce. (3 marks)

Total: 6 marks

Answer

(a) Gradient of line A: $u = -\dfrac{(1.1 - 0)\text{m}}{(0.45 - 0.20)\text{s}} = -4.4\,\text{m s}^{-1}$ ✓

Gradient of line B: $v = \dfrac{(1.1 - 0)\,\text{m}}{(0.80 - 0.50)\,\text{s}} = 3.7\text{m s}^{-1}$ ✓

Gradients with opposite signs ✓

(b) **(i)** impulse of floor on ball = change of momentum of ball

$\Delta p = 0.400\,\text{kg} \times +3.7\,\text{m s}^{-1} - (0.400\,\text{kg} \times 4.4\,\text{m s}^{-1})$ ✓

$= +3.24\,\text{kg m s}^{-1}$ ✓

(ii) average force $= \dfrac{\text{change in momentum}}{\text{time}} = \dfrac{+3.24\,\text{kg m s}^{-1}}{0.05\text{s}} = +64\,\text{N}$ ✓

Ⓔ The positive sign indicates that the floor will exert an upward force on the ball.

Question 10

A student is required to perform an experiment to investigate the properties of a solar cell. His aim is to find the values of the electromotive force (e.m.f.) and the internal resistance of the cell. He also plans to investigate how the power output of the cell depends on the load (external resistance) connected to the cell.

ⓔ This question is based on **core practical 3**. You should be aware of practical techniques used when you performed the experiment.

The circuit he used is shown in the diagram:

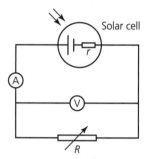

The student used a bench lamp to illuminate the cell and adjusted the resistance box to give an external resistance of 0.50 Ω. He then read the values of the current and the potential difference using the ammeter and voltmeter. The readings were then repeated for a range of values of external resistance.

His results are shown in the table.

R/Ω	I/mA	V/V	P/mW
0.50	374	0.19	
1.00	300	0.30	
1.50	254	0.38	
2.00	225	0.45	
2.50	200	0.50	
3.00	181	0.54	
3.50	163	0.57	
4.00	150	0.60	

(a) State which quantity is the *independent* variable, which are the *dependent* variables and which is a *controlled* variable (3 marks)

(b) State one precaution that should be taken to ensure that the experiment is performed safely. (1 mark)

The student plotted a graph of potential difference against current as shown.

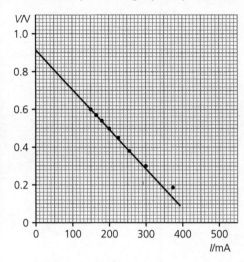

(c) Use the graph to determine the values of the internal resistance and the e.m.f. of the cell.   (3 marks)

🄔 You will need to use the equation $\varepsilon = I(R + r)$ to get an expression for I, V, ε and r in the form $y = mx + C$.

(d) Give one reason why the two plots for the higher values of current do not lie on the same straight line as the rest of the plots.   (1 mark)

(e) (i) Copy and complete the table by calculating the values of the power output of the cell and plot a graph of power against resistance.   (5 marks)

   (ii) Use your graph to find the value of resistance at which the output power of the cell is a maximum.   (1 mark)

   (iii) How could the value of the peak position be determined more precisely?   (1 mark)

   (iv) Write a conclusion on how the maximum power output of the cell depends on the value of the load resistance   (1 mark)

Total: 16 marks

Answer

(a) The external resistance is the quantity varied by the student and so is the independent variable. ✓ The current and potential difference change depending on the value of the chosen resistance and so are dependent variables. ✓ The illumination provided by the lamp needs to be the same throughout the experiment and so is a controlled variable. ✓

(b) Lamps get hot, so avoid contact; the lamp and other equipment should be positioned securely; safety glasses to be worn. [Any one of these ✓]

(c) $\varepsilon = I(R + r) \rightarrow V = -rI + \varepsilon$

So a graph of V against I will be a straight line with a gradient of $-r$ and the intercept on the y-axis will be ε. ✓

Gradient $= \dfrac{(0.90 - 0.20)\,\mathrm{V}}{-0.340\mathrm{A}} = -2.1\,\Omega$ so $r = 2.1\,\Omega$ ✓

Intercept $= 0.90\mathrm{V}$ so $\varepsilon = 0.90\mathrm{V}$ ✓

(d) At higher values of I the internal resistance of the cell gets less (accept changes). ✓

ⓔ The cell is constructed using semiconducting materials. It is likely that the higher currents heat up the cell creating more charge carriers, and so reducing the internal resistance (and decreasing the gradient of the graph).

(e) (i)

P/mW	71	90	97	101	100	98	93	90 ✓

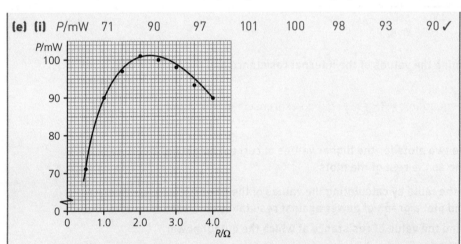

Axes labelled, with units, ✓ suitable scale (covering more than half of available space), ✓ all plots correct, ✓ curve of best fit drawn ✓

(ii) Resistance at maximum power = 2.0 – 2.2 Ω. ✓

(iii) To judge the position of the peak more precisely, more values of P should be found for resistances in the region 1.5 – 2.5 Ω. ✓

(iv) Conclusion: the maximum power output of the cell occurs when the external resistance (load) equals the internal resistance of the cell. ✓

Question 11

ⓔ This is an example of a synoptic question. These can contain questions from any section of the course but, in this case, only parts of core physics I and II and the contents of this guide are used.

An electric car has a mass of 1600 kg. When the car was travelling along a straight, level road with a speed of 30 m s⁻¹, the power was switched off and the car was allowed to 'coast' in a straight line along the road. After free-wheeling for 1.0 kilometre its speed had decreased to 20 m s⁻¹.

(a) Calculate the average acceleration of the car during this time. (1 mark)

(b) Calculate the average resistive force experienced by the car. (1 mark)

The car is fitted with a 144 V battery that powers an electric motor. When driven at a constant speed of 25 m s⁻¹, along the same stretch of road, the current flowing through the motor is 80 A.

(c) Calculate the power needed for the car to maintain this speed. (1 mark)

(d) Determine the efficiency of the motor during this journey. (2 marks)

ⓔ You will need to determine the electrical power developed by the motor.

The manufacturer quotes the average energy consumption of the car as 20 kWh per 100 km.

(e) (i) Convert 1 kWh into the appropriate SI unit. (1 mark)

(ii) Show that the energy consumption of the vehicle at the period of time described above is about 13 kWh per 100 km. (2 marks)

(iii) Give a reason why the calculated value is less than the stated value. (1 mark)

ⓔ You should be aware that the kilowatt-hour (kWh) is a unit of energy.

A simple electric motor can be made using a single coil of wire between the poles of a magnet as shown here. The coil has sides of length 5.0 cm and width of 4.0 cm and the magnetic field strength (flux density) is 600 mT.

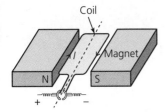

(f) (i) Calculate the maximum torque acting on the coil when a current of 2.5 A flows in the coil. (2 marks)

ⓔ 'Torque' is the technical name for a couple — the moment created by a pair of forces acting on the coil.

(ii) State whether the coil will rotate in a clockwise or anticlockwise manner. (1 mark)

The motors in an electric car need to provide very large torques.

(g) Give two features of an electric car motor that allow it to produce torques that are higher than those created by the single coil motor described in part (f). (2 marks)

Electric cars are able to improve their efficiency by using regenerative braking. The motor function is reversed so that it becomes a generator and recharges the battery.

(h) (i) Use Lenz's law to explain how changing the function of a motor to a generator creates a braking effect. (3 marks)

(ii) Describe the energy transformations that take place during regenerative braking. (3 marks)

(iii) Calculate the energy transferred to the battery when the motor is used to slow the car down from $25\,\text{m s}^{-1}$ to $15\,\text{m s}^{-1}$ (2 marks)

(iv) Give *two* advantages of regenerative braking over conventional systems. (2 marks)

At high speeds air resistance has a significant effect on a vehicle's motion. The diagram shows a large van and a sports car travelling at the same speed along a motorway.

The resistive forces are lower if the airflow relative to the vehicle is laminar, and higher if the flow is turbulent.

(j) **(i)** Explain what is meant by 'laminar flow' and 'turbulent flow'. (2 marks)

(ii) Draw lines around the surfaces of the lorry and the car to represent the airflow, labelling regions of laminar flow and turbulent flow. (2 marks)

Total: 28 marks

Answer

(a) $v^2 = u^2 + 2as \Rightarrow a = \dfrac{v^2 - u^2}{2s} = \dfrac{(20\,\text{ms}^{-1})^2 - (30\,\text{ms}^{-1})^2}{2 \times 1000\,\text{m}} = -0.25\,\text{m s}^{-2}$

$a = -0.25\,\text{m s}^{-2}$ ✓

---

e You must include the minus sign — it indicates that the velocity is decreasing.

(b) $F = ma = 1600\,\text{kg} \times -0.25\,\text{m s}^{-2} = -400\,\text{N}$ ✓

(c) $P = F \times v = 400\,\text{N} \times 25\,\text{m s}^{-1} = 10\,000\,\text{W}$ (10 kW) ✓

(d) electrical power input, $P_{\text{in}} = IV = 80\,\text{A} \times 144\,\text{V} = 11\,520\,\text{W}$ ✓

efficiency $= \dfrac{P_{\text{out}}}{P_{\text{in}}} \times 100\% = \dfrac{10.0\,\text{kW}}{11.5\,\text{kW}} \times 100\% = 87\%$ ✓

(e) **(i)** $1\,\text{kWh} = 1000\,\text{W} \times 3600\,\text{s} = 3.6 \times 10^6\,\text{J}$ ✓

(ii) time to travel $100\,\text{km} = \dfrac{\text{distance}}{\text{speed}} = \dfrac{100 \times 10^3\,\text{m}}{25\,\text{ms}^{-1}} = 4000\,\text{s} = 1.1$ hours ✓

energy per $100\,\text{km} = 11.5\,\text{kW} \times 1.1\,\text{h} = 12.7\,\text{kWh} \approx 13\,\text{kWh}$ ✓

(iii) The car uses more energy when accelerating, climbing hills or travelling along bumpy roads etc. than when moving at a constant speed along a level road. ✓

(f) **(i)** $F_{\text{max}} = BIl = 0.600\,\text{T} \times 2.5\,\text{A} \times 0.050\,\text{m} = 0.075\,\text{N}$ ✓

torque $= F \times$ perpendicular separation $= 0.075\,\text{N} \times 0.040\,\text{m} = 3.0 \times 10^{-3}\,\text{N m}$ ✓

(ii) (Using Fleming's left-hand rule) the coil will rotate anticlockwise. ✓

(g) Any two from: bigger coil; higher current; stronger magnet; more turns on coil; more coils; curved poles; soft iron core (armature) ✓✓

(h) (i) Lenz's law states that the induced current always flows in such a direction as to oppose the change producing it. ✓
When the coils of the motor rotate in the magnetic field an e.m.f is generated across the ends of the coil. ✓
This creates an induced current that opposes the motion and so acts like a brake. ✓

(ii) Kinetic energy ✓ is transferred to electrical energy ✓ and then to chemical energy ✓ in the battery.

(iii) kinetic energy transferred $= \frac{1}{2}mu^2 - \frac{1}{2}mv^2$

$$= \frac{1}{2} \times 1600\,\text{kg} \,[(25\,\text{m}\,\text{s}^{-1})^2 - (15\,\text{m}\,\text{s}^{-1})^2] \checkmark$$

$$= 3.2 \times 10^5\,\text{J}\ (320\,\text{kJ}) \checkmark$$

(iv) Regenerative braking
- increases efficiency (i.e. car uses less fuel)
- reduces wear on brake pads (brakes)

(j) (i) Laminar flow occurs when the paths of adjacent particles do not cross, or if there is no change in the velocity of the particles. ✓ Turbulence occurs when the paths of the particles overlap and eddy currents are produced. ✓

(ii)

Streamlines shown with arrows, ✓ regions of laminar flow and turbulence correctly labelled ✓

ⓔ The diagrams are not definitive. A clear indication that there will be much more turbulence around the lorry is needed.

Knowledge check answers

1 $0.050\,\text{kg} \times 40\,\text{m s}^{-1} = 2.0\,\text{kg m s}^{-1}$

2 impulse $= [0.20\,\text{kg} \times (-8.0\,\text{m s}^{-1})] - [0.20\,\text{kg} \times (+12\,\text{m s}^{-1})] = -4.0\,\text{kg m s}^{-1}$, i.e. upwards

3 $v^2 = 0 + 2 \times 9.8\,\text{m s}^{-2} \times 1.0\,\text{m} \Rightarrow v = 4.4\,\text{m s}^{-1}$

$$F = \frac{\Delta p}{\Delta t} = \frac{0.2 \times 10^{-3}\,\text{kg} \times 4.4\,\text{ms}^{-1}}{80 \times 10^{-3}\,\text{s}} = 1.1 \times 10^{-2}\,\text{N}$$

4 momentum before collision = momentum after collision

$(2.5\,\text{kg} \times 2.2\,\text{m s}^{-1}) + (1.5\,\text{kg} \times -3.6\,\text{m s}^{-1}) = 4.0\,\text{kg} \times v \Rightarrow v = +0.025\,\text{m s}^{-1}$ in direction of 2.5 kg truck

5 E_k before collision $= \frac{1}{2} \times 0.40\,\text{kg} \times (2.0\,\text{m s}^{-1})^2 = 0.80\,\text{J}$

E_k after collision $= \frac{1}{2} \times 0.40\,\text{kg} \times (1.0\,\text{m s}^{-1})^2 + \frac{1}{2} \times 0.20\,\text{kg} \times (2.0\,\text{m s}^{-1})^2 = 0.60\,\text{J}$

Not elastic; kinetic energy has not been conserved

6 a $\omega = \dfrac{\Delta\theta}{\Delta t} = \dfrac{2.0\,\text{rad}}{0.25\,\text{s}} = 8.0\,\text{rad s}^{-1}$

   b $T = \dfrac{2\pi}{\omega} = \dfrac{2\pi}{8.0\,\text{rad s}^{-1}} = 0.79\,\text{s}$

   c $f = \dfrac{1}{T} = \dfrac{1}{0.79\,\text{s}} = 1.3\,\text{Hz}$

7 $a = \omega^2 r = \dfrac{4\pi^2}{(28 \times 24 \times 60 \times 60\,\text{s})^2} \times 4.0 \times 10^8\,\text{m} = 2.7 \times 10^{-3}\,\text{m s}^{-2}$

8 $V = \text{J C}^{-1} = \text{N m C}^{-1}$ $V\,\text{m}^{-1} = (\text{N m C}^{-1})\,\text{m}^{-1} = \text{N C}^{-1}$

9 $\dfrac{F_1}{F_2} = \left(\dfrac{x_2}{x_1}\right)^2 \rightarrow F_2 = \left(\dfrac{4 \times 10^{-2}\,\text{m}}{12 \times 10^{-2}\,\text{m}}\right)^2 \times 1.8 \times 10^{-4}\,\text{N} = 2.0 \times 10^{-5}\,\text{N}$

10 $\dfrac{q^2}{4\pi\varepsilon_0 r^2} = \dfrac{mv^2}{r} \rightarrow v = \sqrt{\dfrac{(1.6 \times 10^{-19}\,\text{C})^2}{9.1 \times 10^{-31}\,\text{kg} \times 4\pi \times 8.85 \times 10^{-12}\,\text{F m}^{-1} \times 0.11 \times 10^{-9}\,\text{m}}} = 1.5 \times 10^6\,\text{m s}^{-1}$

11 a $E = \dfrac{Q}{4\pi\varepsilon_0 r^2} = \dfrac{500 \times 10^{-12}\,\text{C}}{4\pi \times 8.85 \times 10^{-12}\,\text{F m}^{-1} \times (0.25\,\text{m})^2} = 72\,\text{N C}^{-1}$

   b $V = \dfrac{Q}{4\pi\varepsilon_0 r} = \dfrac{500 \times 10^{-12}\,\text{C}}{4\pi \times 8.85 \times 10^{-12}\,\text{F m}^{-1} \times 0.25\,\text{m}} = 18\,\text{V}$

12 $RC = \dfrac{V}{I} \times \dfrac{Q}{V} = \dfrac{Q}{I} \rightarrow \text{unit:} \dfrac{\text{C}}{\text{C s}^{-1}} = \text{s}$

13 $F = BIl = 4.0 \times 10^{-4}\,\text{T} \times 1.2\,\text{A} \times 0.60\,\text{m} = 2.9 \times 10^{-4}\,\text{N}$

14 If the rod was part of a complete electric circuit, a current would flow from west to east (Fleming's right-hand rule, so the west end of the rod is positive and the east end is negative.

15 a $V_0 = V_{rms}\sqrt{2} = 230\sqrt{2}\,V = 325\,V$

b $I_{rms} = \dfrac{I_0}{\sqrt{2}} = \dfrac{5.0\,A}{\sqrt{2}} = 3.5\,A$

16 Carbon-14: six protons and eight neutrons; uranium-235: 92 protons and 143 neutrons

17 $\dfrac{1}{2}mv^2 = Vq \rightarrow v = \sqrt{\dfrac{(2 \times 10 \times 10^3\,V) \times (3.2 \times 10^{-19}\,C)}{6.7 \times 10^{-27}\,kg}} = 9.8 \times 10^5\,m\,s^{-1}$

18 $\dfrac{mv^2}{r} = Bqv \rightarrow v = \dfrac{Bqr}{m} = \dfrac{(1.2 \times 10^{-3}\,T) \times (1.6 \times 10^{-19}\,C) \times (5.0 \times 10^{-2}\,m)}{1.67 \times 10^{-27}\,kg} = 5.7 \times 10^3\,m\,s^{-1}$

19 $E = mc^2 = (9.11 \times 10^{-31}\,kg) \times (3 \times 10^8\,m\,s^{-1})^2 = 8.2 \times 10^{-14}\,J$

20 $m = \dfrac{(1.2 \times 10^{-29}\,kg) \times (3.0 \times 10^8\,m\,s^{-1})^2}{1.6 \times 10^{-13}\,J\,MeV^{-1}} = 6.8\,\dfrac{MeV}{c^2}$

21 $4.04\,u = 6.0 \times 10^{-10}\,J = 3.8\,GeV$

22 $K^0 = \bar{d}s: \left(+\dfrac{1}{3}, -\dfrac{1}{3}\right)$ or $d\bar{s}: \left(-\dfrac{1}{3}, +\dfrac{1}{3}\right)$

Index

Note: page numbers in **bold** indicate key term definitions.

A

acceleration
  centripetal 14–15
  of particles 37–41
alpha-particle scattering experiment 35–36
alternating current (AC) **33–34**
alternating voltage 39, 40
angular displacement **13**
angular speed **13**, 14
annihilation of matter and antimatter 44
antimatter 44
antiparticles **44**, 45, 46
atomic structure 35–37

B

bar magnet, field around 30
baryons 45, 46
braking systems, electromagnetic induction 33
bubble tanks, particle detection 41, 42

C

capacitance (C) **22**
  capacitors 22
  charge and discharge of capacitors 24–26
  energy stored in capacitors 23–24
  measurement of charge 22–23
capacitors 22
  charge and discharge of 24–26
  energy stored in 23–24
centripetal acceleration 14–15
centripetal force 14, 15, 19
charge
  conservation of 42, 45
  measurement of 22–23
circular motion 13–15
coil, magnetic field around 30
collisions 10–13
command terms, exams 48–49
conservation of energy **10–11**, 42

conservation laws
  in creation of electron–positron pair 44
  and particle interactions 45–46
conservation of linear momentum **8–10**
core practicals
  capacitor charge/discharge 25
  collisions 10
  force and momentum 7
coulomb meter 22–23
Coulomb's law **17–19**
current 27–28, 30–32
  alternating 33–34
cyclotrons 39–40

D

deflection of charged particles in magnetic fields 28–29
detection of particles 41–42
direct current (DC) **33**

E

elastic collisions 11, 12
electric fields **16–22**
  Coulomb's law 17–19
  electric field strength 17
  electric potential 19–20
  uniform electric fields 21–22
electric field strength 17
  around a point charge 16, 19
  between charged plates 21
electricity generation 33
electric potential **19–20**
electromagnetic induction 31–33
electron gun 37–38
electrons 19
  deflection of 28–29
  discovery of 37
  rest mass 43
elementary particles 45
e.m.f. 31, 32
energy
  see also kinetic energy

conservation of 10–11, 42
  mass–energy equivalence 42
equipotentials **19–20**

F

Faraday's law of electromagnetic induction **31–32**
fields see electric fields; magnetic fields
Fleming's left-hand rule 27, 28, 29, 39, 41
Fleming's right-hand rule **31**
flux density (magnetic field strength) 27–28, 30
flux linkage **30**
  change in for induced e.m.f. 31, 32
frequency
  alternating current **33**
  circular motion **13**
fundamental particles 45

H

hadrons 45

I

impulse **6**
induced e.m.f. (ε) 31, 32, 33
inelastic collisions 11, 12

K

kinetic energy
  elastic and inelastic collisions 11
  electron–positron pairing 44
  and momentum 12–13
  gain in, particle accelerators 38, 39

L

Lenz's law of electromagnetic induction **32**
  role in electromagnetic braking 33
leptons 45, 46
linear accelerators 38–39
linear momentum, conservation of 8–10
lines of force drawings 16

M

magnetic fields **27**
  alternating current 33–34
  deflection of charged particles in 28–29
  electromagnetic induction 31–33
  flux and flux linkage 30
  magnetic field strength 27–28
magnetic field strength 27–28
magnetic flux density 27, 30
magnetic flux (per unit area) **30**
matter, creation and annihilation of 44
mechanics 6–16
  circular motion 13–15
  collisions 10–13
  conservation of linear momentum 8–10
  momentum and impulse 6
  Newton's second law of motion **6–8**
mesons 45, 46
models of the atom 37
  evidence for Rutherford's 35–36
momentum **6**
  conservation of 8–10, 12, 42, 44
  and kinetic energy 12–13

N

neutrinos 45, 46
Newton's second law of motion **6–8**, 14
nuclear atom 35–37
nucleon number 35

P

pair production 44
particle accelerators 37–41
  cyclotron 39–40
  electron gun 37–38
  linear accelerator 38–39
  relativistic effects 41
particle detectors 41–42
particle interactions 42–44
  and the conservation laws 45–46
peak value (of a current) **33**
period
  alternating current **33**
  circular motion **13**
permittivity of free space 18
photons 41, 42, 44, 45
positron 44
proton number 35
protons 19, 41–42, 45
  proton number 35
  rest mass 43

Q

quarks 45
quark–lepton model 45–46

R

radial field around a point charge 16
rest mass 42, 43
root mean square value **34**
Rutherford model of the atom 35, 37
  evidence for 35–36

S

solenoid, magnetic fields around 30
subatomic particles 45
synchrotrons 41

T

test papers 51–83
time constant **25**

U

unified atomic mass unit 43
uniform electric fields 21–22
  drawing field lines 16